C. Held

Promotion of Innovations in Forest based Small and Medium Size Enterprises of Developing Countries.

Zugl.: Freiburg, Univ., Diss, 2004

1. Auflage 2005
Tel.: 02228-493
Fax: 01212-512382426
eMail: nkessel@web.de

Homepage: www.forstbuch.de, www.forestrybooks.com

ISBN: 3-935638-65-5

Druck: www.business-copy.com

Freiburger Schriften zur Forst- und Umweltpolitik

Band 8

Promotion of Innovations in Forest based Small and Medium Size Enterprises of Developing Countries.

An actor-oriented Analysis of the Colombian Bamboo Sector

Christian Held

Verlag Dr. Kessel

Remagen-Oberwinter 2005

Für Cora,

Alina,

meine Eltern

und meine Großmütter

Acknowledgements

This work has been realised within the project "Guadua Bamboo - Research for sustainable Management and Marketing of Bamboo in Costa Rica and Colombia" funded by the 5th framework programme "International Co-operation with Developing Countries" of the European Union. Further valuable institutional support was provided by the "Proyecto Manejo Sostenible de Bosques en Colombia" especially in person of Ivan Dario Moreno. The Universidad Tecnologica de Pereira deserves special thanks for the provision of infrastructure and personal support, i.e. by the staff of the GIS office and the Dean of the faculty Samuel Ospina. Special thanks go to Juan Carlos Camargo and the department for silviculture. I would like to thank the following persons, who have supported me technically and personally during the field research in Colombia and Costa Rica: Ivan Dario Manzano for being my partner and friend during the conduction of my PhD studies in the Eje Cafetero; Señora Ximena Londoño Pava for supporting me during my stay in Colombia, hosting me and being a personal mentor and friend. Dr. Michael Tistl for guiding me and challenging the quality of my research as well as introducing me to Colombian reality. Jörg Stamm for showing me the potentials of using Guadua bamboo, and for his energetic personality, which has saved me many times from becoming homesick. In Costa Rica I would like to thank Dr. Dietmar Stoian for hosting me at CATIE and Efrain Deras for taking me to the Caribbean Sea. In Ecuador I deeply thank Dr. Arienne Henkemans and Dr. Eric Cleuren from INBAR for supporting my research in Ecuador and for their co-operation in publishing my first study results.Very special thanks go to Prof. Michel Becker, who has been an excellent personal and professional supervisor leaving me every space I needed to realise my ideas and complete the present work, and to Dr. Jochen Statz for accompanying my research for two years and providing valuable professional inputs as well being a pole of peace and harmony for my troubled mind.

Christian Held

Freiburg, den 31.12.2004

Table of Contents

List of tables

List of graphics

List of Diagrams

List of Figures

Abbreviations

ANDI	Asociación Nacional de Empresarios de Colombia
ASOCAÑA	Asociación de Cultivadores de Caña de Azúcar de Colombia
AUC	Autodefensas Unidas de Colombia
a.s.l.	above sea level
CAR	Corporación Autonoma Regional
CARDER	Corporación Autonoma Regional de Caldas
CATIE	Centro Agronomico Tropical de Investigación y Enseñanza
CENICAFE	Centro Nacional de Investigaciones de Café
CENICAÑA	Centro Nacional de Investigaciones de Caña de Azúcar
CENIPALMA	Centro Nacional de Investigaciones de Palma de Aceite
CIA	Central Intelligence Agency
CINEP	Centro para la Investigación y Educación Popular de Colombia
CIPAV	Centro para la Investigación en Sistemas sostenibles de Producción Agropecuaria
cm	centimetre(s)
CO2	Carbon Dioxide
Col.$	Colombian Pesos
COLCIENCIAS	Instituto Colombiano para el Desarrollo de la Ciencia y la Tecnologia
CONIF	Corporación Nacional de Investigación y Fomento Forestal
CORPEI	Corporación de Promoción de Exportaciones e Inversiones en Ecuador
CORPOCALDAS	Corporación Autonoma Regional de Caldas
CRQ	Corporación Autonoma Regional del Quindio
CVC	Corporación Autonoma Regional del Valle de Cauca
DANE	Departamento Administrativo Nacional de Estadistica
DIAN	Dirección de Impuestos y Aduanas Nacionales de Colombia
E&E	Education, dissemination and extension activities
e.g.	exempli gratia (for example)
EFI	European Forest Research Institute
ELN	Ejercito de la Liberación Nacional
et al.	et alli (and others)

etc.	et cetera
EU	European Union
f.	following page
ff.	following pages
FAO	Food and Agriculture Organisation of the United Nations
FARC	Fuerzas Armadas Revolucionarias de Colombia
FEDECAFE	Federación de Cafeteros de Colombia
FEDEPALMA	Federación Nacional de Cultivadores de Palma de Aceite
FDI	Foreign Direct Investments
FINAGRO	Fondo para la Financiación del Sector Agropecuario
FOMIPYME	Fondo Colombiano para la Modernización y Desarrollo tecnologico de las Micro-, Pequeñas y Medias Empresas
FOREC	Fundación para la Reconstrucción del Eje Cafetero
FUNBAMBÙ	Fundación del bamboo en Costa Rica
GDP	Gross Domestic Product
GFAR	Global Forum on Agricultural Research
GNP	Gross National Product
GTZ	Gesellschaft für Technische Zusammenarbeit (German Development Co-operation)
ha(s)	Hectare(s)
I&I	Promotion of interactions and institutions building
ICONTEC	Instituto Colombiano de Normas Tecnicas y Certificación
i.e.	id est (that means)
IICA	Instituto Inetamericano de Cooperación en Agricultura
IFI	Instituto Fomento Industrial
INBAR	International Network for Bamboo and Rattan
Indep.	independent
IMF	International Monetary Fund
ILO	International Labour Organisation
incl.	including
ITTO	International Timber Trade Organisation
km	kilometre(s)
km^2	square kilometre(s)
m	meter(s)
m^2	square meter(s)
m^3	cubic meter(s)
Mio.	Million
modif.	modified
MSBC	Proyecto Manejo Sostenible de Bosques en Colombia
N	Total number
n	Total number

NGO	Non Governmental Organisation
NIE	New Institutional Economics
NIS	National Innovations System
No.	Number
NTFP	Non Timber Forest Product
N/A	Not available
OECD	Organisation for Economic Co-operation and Development
P&P	Pulp and Paper
PCS	Production-to-Consumption-System
PYMEs	Pequeñas y Medias Empresas
QCA	Qualitative Contents Analysis
R&D	Research and Development
RIS	Regional Innovation System
SENA	Servicio Nacional de Aprendizaje
SCB	Sociedad Colombiana del Bambú
SME	Small and Medium Size Enterprises
SPSS	Statistical Programme for Social Sciences
t	metric ton(s)
Transl.	Translated from
UNESCO	United Nations Educational, Scientific and Cultural Organisation
UNCED	United Nations Conference on Environment and Development
UNIDO	United Nations Industrial Development Organisation
US$	United States Dollar
WCO	World Customs Organisation
WTO	World Trade Organisation
€	Euro

Average Exchange Rates 2002

US$ 1 = € 1.06

US$ 1 = Col. $ 2,580

€ 1 = Col. $ 2,451

1. Introduction

The present study was carried out within the framework of the European Union funded research project "*Guadua-Bamboo. Research for sustainable management and marketing of bamboo in Costa Rica and Colombia*". During the initial phase of the project in early 2002, an analysis of the current state of the Colombian bamboo 'production to consumption system' was conducted, which was published by HELD AND MANZANO in 2004. The outcomes of this first investigation phase were rather daunting when one considers the expectations the actors involved have for the development of this resource base. The status assessment research revealed an economic sector constituted of micro, small and medium size enterprises, which primarily work on a subsistence and highly artisan basis. The enterprises investigated in this study were not organised, and the product variety and quality gave rise to the assumption that the sector is, in its current state, not competitive in either domestic or international markets.

Nevertheless, the Colombian actors at the governmental and departmental level, as well as NGOs and actors from the agricultural sectors, credit the bamboo sector with great potential to contribute significantly to the social and economic development of the rural areas in Colombia, i.e. through the promotion of industrialised production processes. This expectation is rooted in the basic assumption that regional development is, among other things, based on the optimal employment of resources available locally. The value of these resources should be maximised through technological progress and industrialisation, induced by a set of promotional measures ranging from extension and dissemination activities to export and trade promotion.

When the Colombian bamboo sector was examined in 2002, the promotional and organisational processes had already been underway since 1999, some even longer. However, at the first glance the state of the sector's enterprises did not reveal significant progress. As a result, the actors involved, as well as the researchers, began to ask themselves whether the promotion activities conducted were adequate. The question of whether the enterprises in the sector were capable of fulfilling societies` expectations, in terms of technological progress and increasing competitiveness, even with the support of the state, NGOs, technical co-operation and other organisations, also arose. This led to the formulation of the basic research interest of this study: Are the promotion measures applied in the Colombian bamboo sector adequate to bring about technological progress?

In the following sections of this introduction chapter, the foundation for the theoretical discussion is laid out. The place forest resources are allocated in development strategies is discussed, followed by an introduction to the world's bamboo sectors revealing that the basic problems currently encountered by the Colombian bamboo sector are shared by bamboo-based sectors on all continents. A special emphasis is given to the Chinese bamboo sector and its development, as it is perceived as a role model for promotional measures in Colombia, and in other countries. The final section brings together the central research issues at the core of this study.

1.1 Forests in Development

Over the past number of decades the sustainable use and management of natural resources has increasingly gained in importance in both national and international promotion programs and strategies for the development of rural areas in developing countries. A special emphasis has been placed on forest resources[1], which have been in the centre of the international debate on global warming, climate change and carbon fixation since the UNCED in Rio in 1992[2]. The same forests constitute the livelihood of several million people all over the world, through the provision of timber, firewood, non-timber products, meat and agro-forestry production systems. At the same time, the forests represent ecosystems rich in biodiversity, and vital for the protection of local soils and water regimes.

Probably the most important forest resource at this time, in economic terms, is the land on which the forests are situated. This land only becomes available for economic production purposes once the forests themselves have been felled. In developing countries, these production forms are usually small-holder agricultural production systems, or the establishment of large scale cash crops and timber plantations[3]. A further important land use change is the mining of former forestland for all kinds of economically important minerals, and for the exploitation of oil resources[4]. The exploitation of raw materials from the forests is the first step towards the clearance of so called wilderness areas, and the settlement and cultivation by agricultural producers[5]. The decline in forest areas depends greatly on the difference between the opportunity costs of maintenance and those of transformation.

The conclusion that can be drawn from these facts is that a forest protection strategy must be accompanied by an economic valuation of these resources[6]. Hence, current international, national and regional development strategies for the sustainable utilisation of forest resources consider equally social, ecological and economic development aims.

Traditionally, the contribution of natural resources to the development of a region or a nation was seen in terms of their value as exploitable raw materials, such as minerals, timber or foodstuffs[7]. The natural resources and their economic value have been central to a number of

---

[1] Other important natural resources, which constitute part of the same discussion and strategies, are the sweet water resources and the oceans, the management of mountain areas, etc.

[2] PÜTTMANN (2002: IIIff.)

[3] FAO (2001: 44); WILLIAMS (2003)

[4] FINGER (1999)

[5] WILLIAMS (2003)

[6] PÜTTMANN (2002: Vf.)

[7] WUNDER (2003: 13ff.)

theoretical approaches, beginning with the land rent theory[8] and the trade with Ricardo commodities[9], through the extraction economics of natural resources[10] to the theoretical approach of the 'Dutch Disease' and its extended models[11]. The common notion shared by these theoretical approaches is that the economic value of natural resources is limited compared to that of the manufacturing sectors. Ultimately, the sustainable profit-oriented extraction of raw materials will result in higher returns than the transformation of the resource base and exhaustive extraction of natural resources[12].

Forest products, such as medicinal plants, seasonal fruits and foods, and materials, such as bark, fodder, rattan or bamboo, have traditionally been treated as minor forest products with no significant importance to regional and national economies. However, recent studies have shown that these products also contribute significantly to the rural livelihood of the forest dwellers and adjacent communities[13]. The assumption is that the sustainable extraction and use of these minor forest products, or non-timber forest products (NTFPs[14]), as well as the harvesting of timber and other 'major' forest products contribute to a) a higher valuation of the forest resource and, therefore, act as a barrier protecting against the destruction of the resource base[15]; and b) they may constitute an important part of rural incomes and subsistence strategies, and as such are inseparably linked to the maintenance of the forest resource base[16].

To maintain and protect forest resources from the economic needs of society, protection strategies have been put in place. These protection strategies have evolved continuously. The pure protection approaches of the 1960s and '70s excluded the local population from any kind of use and tenure rights. In the 1980s and '90s protection approaches were geared towards integrated co-management by protection entities and the local population of the protected areas. Protected areas were stratified into core, buffer and management zones, an approach ensuring the acceptance of the protection idea by locals, and at the same time granting socially just and economically necessary access and use rights[17]. The final statement of the

---

[8] NAVARRO (2003)

[9] RICARDO (1817)

[10] ENDRES AND QUERNER (1993)

[11] WUNDER (2003)

[12] ENDRES AND QUERNER (1993: 151ff.) and PÜTTMANN (2002: VI)

[13] STATZ (2000b: 1ff.)

[14] According to DEBEER AND MCDERMOT (1989), NTFPs encompass *"...all biological materials other than timber, which are extracted from natural forests for human use"*.

[15] PETERS ET AL. (1989); See the publications by DURST, ULRICH AND KASHIO (1994); PERÉZ, SAYER AND COHEN (1993); TEMU (1995) as early examples of NTFP studies. Although the more recent studies increasingly seek to explain socio-economic parameters and development factors, the theoretical fundaments and methodologies are divergent. For an overview of socio-economic forest resource case studies and comments on theories and methods see PERÉZ AND BYRON (1998).

[16] HYTTINEN, P., MONONEN, A. AND PELLI, P. (1996)

[17] THE WORLD BANK (2001); FAO (2001: 22ff.); PÜTTMANN (2002: VIIff.)

XII Word Forestry Congress emphasised the role these forest resources may play in poverty alleviation, environmental protection and development in general[18].

Both the harvesting of timber and the extraction of NTFPs (either in unprotected forest areas or in the management zones of protected forest areas) only contribute to the maintenance of the forest resources, when continued sustainable use is perceived as more profitable than the transformation to other types of production. In both cases, the value-adding processes taking place determine the ultimate benefits to the users and, consequently, their perceived valuation of the forest resource. The value of a forest in the first line is defined by the direct economic benefit derived from it[19]. Hence, the most effective means of protecting forest areas and maintaining forest resources are to derive the highest possible added value from forest-based products, and distribute the benefits amongst the user groups living off the forests[20].

Today, vast forest areas worldwide are degraded or secondary forests, scattered between agricultural land and settlements[21]. The utilisation pressure upon these forest areas is high and their survival depends on the benefits the rural population can derive from them. These commonly include the provision of firewood and construction timber, and the availability of grazing areas for cattle and other livestock, plus various NTFPs. In these areas, the existence of forest vegetation is crucial to the protection of the soils and the maintenance of the water balance. Furthermore, the forests constitute the only remaining freely available sources of the aforementioned materials and products for local populations. Current rural development strategies emphasise the importance of the recognition of the value of these forest resources from social, ecological and economic development perspectives[22].

Basically, the ecological impact of forest resource utilisation on rural development depends on the pressure exerted upon the resources by local and non-local actors. The pressure placed on the resource is, in turn, dependent upon the social and economic demands upon the rural population. The fulfilment of these demands forms part of the development strategies, which seek to optimise the socio-economic aspects of the development process. The contribution of forest resources to these socio-economic issues is viewed in two ways: 1) the contribution of forest resources to subsistence and day-to-day needs through the extraction of products and services, and their consumption and trade; and 2) the contribution to the regional or national economy by the sectors based on forest resource use, in terms of the GNP, employment figures and trade statistics. In rural regions either one of these two perspectives may dominate, depending on the existing framework conditions. The first perspective usually dominates in regions with poorly developed agriculture and industry, which in turn results in community forestry approaches and the institutional regulation of use and access rights. In

[18] KOYO (2003)

[19] Compare FAO (2001: XIf.)

[20] Compare GTZ (1997); FAO (2001: 13ff.)

[21] According to the FAO forest classification, which refers to the percentage of canopy cover, these forests are fragmented and open forests. FAO (2001)

[22] FAO (2001: 6ff.)

rural regions, which are agriculturally and industrially more advanced, the economic sectors based on forest resource use, and their enterprises, are the focus of the development strategies[23].

The promotion of community forestry approaches draws upon sociological and ethnological theories and approaches. Economic aspects do not play a central role. The common promotion strategies of economic sectors based on forest resource use in developing countries, however, are based upon the basic assumptions of business location, competitive advantage and factor productivity of the neo-classical and new economic growth theories. Central to the promotion of these sectors is the initiation and improvement of value–adding processes through the input/transfer of knowledge, capital and technology. According to the literature related to the promotion strategies of forest sectors in developing countries, the bottleneck is shortcomings in the value-adding capacity of these sectors. This results in a weak exchange factor in international trade relations and low importance within the national economies. The recommendations made to overcome these problems include the classical methods, namely the transfer of technology and knowledge from industrialised countries and the attraction of Foreign Direct Investments (FDIs) to acquire business and technological expertise from private enterprises. In general, these sector promotion measures should benefit the small and medium size enterprises (SMEs), which according to economic theory are the driving force behind employment creation and technological innovation[24].

Although the innovation processes and the acquisition of knowledge and technology by the entrepreneurs and farmers are perceived as crucial to the development of forest based sectors, empirical data mainly exists for the industrialised countries. The development of promotion strategies by international and local entities has been based on these experiences, for example those of the timber based industries of Scandinavia, Germany and North America[25]. However, the societal and micro-economic environment of the developing countries differs significantly from the conditions encountered in the industrialised regions of the world, as do the demands the forests have to satisfy[26]. However, there is useful evidence from innovative research in the agricultural sectors of the developing world, which underlines the necessity to investigate the situations of the individual farmers and entrepreneurs, and their knowledge and access to technology[27].

In summary, a general set of problems arise in relation to development and growth-oriented approaches in the forest-based economic sectors, which affect the management strategies and

---

[23] Compare FAO (2001: 22ff.)

[24] Compare GTZ (1997) and GTZ (2001a)

[25] E.g., BONITA ET AL. (2002); FERRANTI ET AL. (2002)

[26] Compare the different trade statistics of wood-based products in OECD and developing countries in WHITEMAN, BROWN AND BULL (1999).

[27] MELLOR AND DESAI (1985); See also the publications of the FAO agricultural division and agricultural technology transfer, research and development (R&D), and the proceedings and publications of the Global Forum on Agricultural Research (GFAR).

the value-adding processes of the entrepreneurs and farmers. These problems also shape the promotion and regulation measures created and enforced by the actors. The problems are:

- There is a fundamental rivalry between the best protection and the best economic utilisation of forest resources. Hence, one of both is always below the optimum.
- The economic and surrogate values the forests contribute to the rural economies are not fully known and their assessment has not advanced significantly. Hence, the values used in argumentations are based upon estimates and models.
- There is a large discrepancy between the relative importance of the forest resource for the household economies, on the one hand, and the national economies, on the other. Hence, national economic policies usually bypass local users and their management practices.
- The added value on forest products, which constitutes part of the regional, national and international value chains, is comparatively low. Hence, the opportunity costs are high compared to other raw material production and manufacturing opportunities, as are investments.
- The value-adding processes in sectors based on forest resources are one important bottleneck in relation to the increased valuation of the resource, its protection and its contribution to rural development and the economy.
- A high added value is linked to knowledge and technology, which has not yet evolved or has not yet been transferred to the forest product based sectors in developing countries. Hence, value-adding as a driving force for the economic valuation of the resource is not working.
- Strategies to promote forest products are either factor-input-oriented or product-output-oriented. Hence, the endogenous processes and dynamics in these sectors are not subject to promotion strategies.

1.2 Examples from the World's Bamboo Sectors

The aforementioned problems of the forest resource based sectors can typically be observed in the bamboo producing and processing sectors across the globe. A well-documented set of bamboo sector case studies, which have taken place in the past ten years, enables a comparative analysis of the major problems in achieving the respective development goals of these sectors.

In many developing countries the bamboo resources in particular have become increasingly interesting. The reason for this is that bamboo, which according to international classification is a NTFP[28], has proven to contribute successfully to rural development in several provinces

[28] See FAO (2001b: 81ff.).

of the People's Republic of China[29]. The international economic success of the Chinese bamboo sector and its positive consequences for the rural population and economy has encouraged many other countries to promote and utilise their own bamboo resources. There are more than 22 million ha of bamboo stands worldwide[30]. The vast majority of the world's bamboo resources are found in Asia, i.e. in China (4 million ha) and India (10 million ha)[31]. Annually, bamboo worth more than US$ 5 billion is traded internationally[32].

The volumes of bamboo harvested and used for domestic purposes is unknown. In China US$ 800 million are earned from bamboo exports. More than 114 million culms are harvested annually for all kinds of purposes[33]. The 'timber of the poor', as bamboo is commonly termed in many countries, covers the daily needs of hundreds of thousands of people all over the world[34]. In India, for example, there are more than two million traditional bamboo artisans, and in China millions of farmers grow bamboo for use in industrial manufacturing industries[35]. The success of the Chinese bamboo industry is commonly attributed to massive governmental support and the innovativeness of the related industries, which have developed a number of new bamboo products and processing technologies in the recent years[36]. No other country has, as yet, been able to seriously challenge the competitiveness of the Chinese bamboo industry, due to the enormous resource base, the institutional framework, the labour market, the large domestic and regional consumer markets, and the technological head start.

The international network for bamboo and rattan (INBAR), which started its work in the year 1997 with 12 member countries, currently has 28 members from all continents[37]. Many countries, particularly African, have recently joined the INBAR (in 2002 and later). This is an indicator of the increasing expectations these countries have of the economic development potential of this resource. In most countries bamboo currently contributes mainly to the livelihoods of local populations and to small scale economic sectors, such as tooth pick manufacture, handicraft, furniture and basket weaving[38]. The INBAR agenda fixed the socio-

---

29 VANTOMME (2003: 43)

30 JIAFU (2001: 1)

31 BELCHER (1995: 2f.)

32 This figure is based on the INBAR estimates for the year 2003, as no statistical sources for the total volumes of internationally traded bamboo products are available. INBAR director Ian Hunter provided this figure during a discussion on the internet discussion platform 'Bamboo-plantations' on the 28th of April 2003. In the year 2002 INBAR and the ITTO initiated the development and introduction of new international custom codes for bamboo products for an improved registration and documentation of the international bamboo trade (see WARDLE, 2002). The World Customs Organization (WCO) will officially introduce these new codes into the harmonised system for international customs codes in 2007.

33 NON-WOOD NEWS (2003)

34 JANSSEN (1998)

35 BELCHER (1995: 3ff.)

36 Non-wood News (2004: 25ff.)

37 Canada is the only member of INBAR, which does not possess own natural bamboo resources. The other member states are made up of ten countries from Asia, ten from Africa and seven from Latin America

38 INBAR has funded and prepared a number of national case studies, such as ADAMSON AND LÓPEZ (2000); ADKOLI (2002); ONGUGO ET AL. (2000), SINGH, BERSALONA AND QUINTANS (2000); ZHAOHUA (2001); etc.

economic importance of the bamboo resources in a plan of action for the economic development of the resource. These activities comprise[39]:

- The design and transfer of efficient low-cost technology
- The expansion of the applications of bamboo and rattan resources
- The creation of incentives for the regeneration of natural stands
- The calculation of the profitability of alternative marketing and production systems
- Assessing the users` needs for potential interventions
- The creation of structures for the conservation of the resource
- The promotion and intensification of bamboo and rattan cultivation
- The examination of existing policies and the recommendation of policy reforms

The success story of the Chinese bamboo sector is certainly linked to the long cultural tradition of bamboo and its use in China. However, such traditions exist in many other countries also. Additionally, Chinese forest policy has also had a positive influence on the bamboo sector, favouring the plantation and management of bamboo stands to substitute and protect the national timber and forest resources[40]. Currently in China there are more than 4 million ha of bamboo stands (natural and plantations) and another 70 000 ha are planted annually[41]. Almost 1 000 factories and enterprises currently depend on the supply of bamboo[42]. More than 500 000 farmers have established bamboo plantations in the six major bamboo producing and manufacturing Chinese counties in the recent years[43].

In China bamboo products make up 55% of the gross annual agricultural output value[44]. For many countries seeking to promote their domestic bamboo sectors, the Chinese experience represents a role model to be followed. However, it is important to note that the Chinese economic and political conditions differ significantly from those of most other countries. Apart from the international differences between the countries, internal factors may arise, which make the transfer of successful experiences and lessons difficult. PÉREZ ET AL. have shown how important the structural environment is for the further development of the bamboo resource[45]. The degree of industrialisation and the development of the secondary and tertiary sector in a region, in particular, influence the development potential of the bamboo sector. Less developed regions tend to benefit more from the development of the bamboo sectors than more advanced regions.

---

[39] As in INBAR (1999: 2ff.)

[40] A detailed description of the Chinese forest sector policies is given in ZHANG ET AL. (1999). For the effects of the Chinese forest and economic policies on the bamboo sector see KANT AND CHIU (2000) and PÉREZ ET AL. (2001)

[41] NON-WOOD NEWS (2003)

[42] WEI AND CHI-SON (1997)

[43] PÉREZ ET AL (2001: 10)

[44] JIAFU (2001: 4)

[45] PEREZ ET AL. (1999 and 2001)

In Latin America bamboo is a common but underestimated and underutilised resource[46]. It has only been in recent years that public and private actors have put more emphasis on the development and promotion of the regional bamboo resources, as a result of the economic and agricultural crisis to hit the continent. Colombia and Ecuador are currently very active in promoting their respective bamboo sectors and actors. Ecuador houses the 'International Network for Bamboo and Rattan' (INBAR) office for Latin America, which was opened in 2001. Ecuador is an important exporter of raw bamboo to neighbouring countries, for example Chile[47]. However, since the 'dollarization' of the country the export activities have decreased significantly[48]. Meanwhile, a number of private enterprises have established production sites for high-value bamboo products, such as flooring elements and panels. These products have been successfully traded on international markets by Chinese and Asian enterprises for over ten years. The main potential export market for Ecuadorian bamboo product manufacturers will be the U.S.[49].

In Colombia bamboo and bamboo products have mainly been utilised in the construction of houses, and for auxiliary uses in agricultural production systems[50]. The major Colombian bamboo producing and processing region is the Andean coffee growing region. During the colonisation of the region almost all the natural forests and large areas of bamboo stands were transformed into arable land[51]. Only residues of the natural forests have survived, in places where agricultural cultivation was not deemed economically feasible, and at altitudes above the optimum growth conditions for coffee. In contrast to the forest resources, vast bamboo resources have survived and were even actively maintained and managed in the midst's of cultivated areas. Today between 30 000 ha[52] and 40 000 ha of bamboo can still be found in Colombia. In the coffee growing region of Colombia, bamboo is part of the culture and constitutes the livelihood for a number of farmers, intermediaries and enterprises[53].

Brazil is probably the most important Latin American country in terms of bamboo resources and markets for bamboo products. However, with the exception of some basic data and information on the bamboo species and the resource base[54], bamboo related information is scarce and mainly based on expert estimations. Brazil is not a member of INBAR and the

---

[46] MORÁN (2001)

[47] In Chile the bamboo species *Guadua angustifolia*, which is the most common species traded in Ecuador, is known as '*Caña Guayaquil*', which means the cane from the Ecuadorian city Guayaquil.

[48] CORPEI (2002b and 2002c)

[49] See the studies by DAGILIS AND WIT (2003), LOAIZA, MONTENEGRO AND OLAYA (2001) and CRESPO & ASSOCIADOS (2001)

[50] For an overview of the uses of bamboo in Colombia and other countries see HIDALGO, 2003.

[51] BERNAL (2002: 58f.). According to latest site classifications, more than 40 000 km² in the MSBC-project pilot region could be suited to bamboo (*Guadua angustifolia*) plantations, which includes the Eje Cafetero plus the department of Tolima (total area of the pilot region: 58 000 km²) (BERNAL, 2002: 10f.). This indicates that before the colonisation this area, at least, could have been stocked with bamboo and mixed bamboo-forest stands.

[52] GIRALDO AND SOBOGAL (1999: 35)

[53] HELD AND MANZANO (2004)

[54] E.g., in LONDOÑO (2001)

domestic bamboo sectors receive little public interest and funds. The most important national actor is the 'Bamboo Brazil Foundation' NGO. In Costa Rica, which like Brazil is not a member of INBAR, the government implemented a national bamboo programme[55] in the late 1980s in support of the domestic bamboo sector. This support culminated in the creation of the national bamboo foundation, FUNBAMBÙ[56], which focuses mainly on the propagation and construction of social housing using bamboo[57].

DERAS (2004) recently analysed the Costa Rican bamboo sector. The study reports that the production of Costa Rican bamboo products has doubled within the last ten years[58], mainly handicrafts for the tourist and domestic market. For other Latin American countries, such studies and information are practically non-existent. Compared to the rest of Latin America, the Colombian and Costa Rican bamboo sectors are probably the most active and advanced, in terms of traditional appreciation, the intensity of research activities and contemporary importance in the national political agenda. In *table 1* the most important case studies of national bamboo sectors are listed and compiled in groups according to their respective continents.

Table 1. List of national bamboo sector case studies according to continent

Continent	Case study authors
Asia	ADKOLI (2002), JIAFU (2001), PEREZ ET AL. (2001), KANT AND CHIU (2000), SINGH, BERSALONA AND QUINTANS (2000); PEREZ ET AL. (1999),MAOGONG ET AL. (1998); KARKI, SHERCHAN AND KARKI (1998); SRIVASTAVA (1998); MATHEW (1998); WEI AND CHI-SON (1997), HANUMAPPA (1996)
Africa	ESEGU, SSENTEZA AND SEKATUBA (2000); KELBESSA ET AL. (2000); ONGUGO ET AL. (2000); TAFORI (2000)
Latin America	HELD AND MANZANO (2004), DERAS (2004), DAGILIS AND WIT (2003), LOAIZA, MONTENEGRO AND OLAYA (2001), CRESPO & ASSOCIADOS (2001), ADAMSON AND LÓPEZ (2000)

1.3 Aim of the Research

Over the past number of decades, the sustainable use of forest resources and value-adding processing of forest products have increasingly been identified as possible means of sustainable rural development. The value adding-processes have been identified as a major bottleneck, as technological processing facilities are lacking, and the state of technical and entrepreneurial knowledge in the forestry sectors of developing countries is generally low. Thus, the developing countries serve rather as the suppliers of raw forest materials, with no significant value-adding taking place. A large number of micro, small and medium-size

[55] *Proyecto Nacional de Bambú*

[56] *Fundación del Bambú en Costa Rica*

[57] ADAMSON AND LÓPEZ (2000)

[58] DERAS (2004: 101f.)

enterprises (SMEs) active in forest and timber based sectors have developed value-adding processes according to their entrepreneurial, technical and financial capacities.

The development strategies in place today seek to resolve these problems, mainly through education, extension and dissemination measures. These activities will encourage investment and innovation processes in the forest based enterprises, enhancing value-adding processes and improving competitiveness in domestic and international markets. The promotion strategies applied draw from the experiences made in the forestry and timber processing sectors of industrialised countries. The reality of innovation processes in forest-based enterprises in developing countries has hardly been subject to systematic research. The same is true of the effectiveness of the promotion strategies.

This study investigates these open questions, placing an emphasis upon the innovation processes in bamboo-based SMEs, and the promotion strategies applied to enhance these processes by institutional actors. The study has been carried out from a socio-economic perspective, assuming that the actors investigated, the entrepreneurs and the institutional actors, are driven by patterns of profit maximisation, as well as by the social and institutional structures. Both economic and socio-institutional aspects define the actors' 'bounded rationality' and their patterns of behaviour. The basic questions are: 'How do these entrepreneurs innovate?' and 'How effective are promotion strategies in enhancing these processes?'. The results of this study will contribute to 1) the theoretical body of knowledge on innovations and technological progress, 2) the improved anticipation of forest-based sector promotion strategies, and 3) the extension of the base of empirical knowledge relating to Latin American bamboo sectors and their economic development.

1.4 Structure of the Study

Beginning with the problems and deficits facing forest resource based sectors in developing countries, an investigation of the existing theoretical work relating to the interactions between development, economic growth and natural resources is carried out. In the *chapter 2*, the history of development theory is outlined, in particular the modernisation theories and the dependency theory. The *chapters 2.1 and 2.2* discuss the different understandings of economic sectors in the context of economic growth and regional development. The goal is to gain an insight into the current understanding of the relationship between growth and development. The role forest resources have played in historical development and growth processes is highlighted in *chapter 2.3*.

In *chapter 3.1* the sources of economic sector competitiveness are analysed. The basic competitive advantage approach is advanced to that of systemic competitiveness, which emphasises the importance of the actors and their interactions for sector development and competitiveness. As the driving force behind growth and development, technological progress and innovations are identified in *chapter 3.2*. This aspect is further discussed in greater detail in *chapters 3.2.1 – 3.2.4*. The theoretical discussion continues with an introduction to the mentality and understanding of the actors, and their behaviour as it relates to this study (*chapter 3.3*). The *chapter 3.4* concludes with a list of explanation deficits encountered in the discussed theories, which lead to the research guiding hypotheses.

The *chapter 4* introduces the reader to the study region. The region is a role model for many other regions in developing countries seeking to enlarge and diversify their economic activities and orientation, by focusing on the available natural resources. The study region was located in the bamboo producing and manufacturing sector of the Colombian coffee growing region. In *chapter 5* the research methods (*chapters 5.1* and *5.2*) and analytical tools (*chapter 5.3*) applied are introduced and briefly discussed in *chapter 5.4*.

Chapters 6 - 8 present the results of the research. *Chapter 6* focuses on the meso-level actors, their strategies and interactions, and the outcomes of an institutional and framework analysis of the study sector. In the following chapter (*chapter 7*), the meso-level actors' strategies for the promotion of the Colombian bamboo sector's enterprises are analysed. The final results chapter (*chapter 8*) reveals the innovativeness of the sixty small and medium sized enterprises and the six large-scale raw material producers investigated. The results of these three chapters are summarised in the first chapter of the discussion (*chapter 9.1*). This is followed by the discussion of the core aspects of the study; the actors and their interactions, and the consequences these interactions have for innovation in the study sector (*chapter 9.2ff.*). Finally, conclusions are drawn for theory and further research.

2. Theories of Development and Growth

The relationship between development and growth in development and growth processes of developing countries has been discussed for many decades now. Development itself is a most complex term and difficult to define[59]. Basically, it is a process of change from one status to another, which is usually perceived as an improvement[60]. This very abstract description of the process implies that development can be measured by using a set of parameters to compare the different phases during the development process. In the context of societal development the changes can be of different natures, e.g. economic changes, changes in the political system, the quality of life or the reduction of poverty. These parameters usually depend on the perceptions, values and attitudes of political actors, which determine how to design development strategies[61].

A number of theories have been developed to attempt to explain the preconditions for development and the reasons for underdevelopment. The most important theories relate to the 'modernisation' and 'dependency' approaches. The 'modernisation theories' are based on neo-classical endogenous growth theories[62] whereas the 'dependency theory' is founded on the assumption of structural inequality, as represented by economic or political centre-periphery models[63]. These theories dominated the debate on development and growth in the 1970s and 1980s.

In the 1950s and 1960s the majority of the world's developing countries began to adopt development-oriented policies and strategies aimed at economic growth, to catch up with the industrialised nations[64]. These strategies under the paradigm of "development through growth" were modified and applied until the 1980s[65]. Until the 1970s development was basically understood as an equivalent to economic growth and development strategies aimed at the creation of favourable economic conditions (the basic features of economic growth and development are compared in *table 2*). They were thought to enable underdeveloped countries to imitate economic, and therefore societal, models of the industrialised countries and to catch up with them[66]. This concept of development and development strategies is based on the neo-classical growth theory and assumes that growth is basically a result of the increased

---

[59] See KÖSSLER (1998: 23) and NOHLEN AND NUSCHELER (1993a: 56)

[60] For a detailed discussion of the term 'development' see STATZ (2000: 21ff.).

[61] The creation of such measurable parameters leads to the discussion of development indicators, and was discussed by NOHLEN AND NUSCHELER (1993b: 76ff.).

[62] As proposed by SOLOW (1956 and 1970), for example. The most widely known and adopted neo-classical growth model was that of SOLOW, which has since be named the SOLOW Growth-Model (ROMER, 2001: 7).

[63] Originally based on the theory of imperialism as proposed by Hobson or Lenin (THIEL, 1999: 11).

[64] MENZEL (1993: 133ff.)

[65] See THIEL (1999: 9ff.) and NUSCHELER (1993: 156ff.)

[66] NOHLEN AND NUSCHELER (1993a: 59)

productivity of the capital[67]. Capital accumulation and net-investment combined with an increasing labour market will result in economic growth. Therefore, the promotion of entrepreneurs and industries has priority over the redistribution of capital resources within society[68].

Table 2. Basic features of development[69] and economic growth[70]

Feature	Economic growth	Development
Growth	Mainly quantitative	Quantitative and qualitative
Strategy	Capital-oriented	Labour-oriented
Distribution of added-value	Economic investment	Social redistribution
Progress	Technological innovation	Knowledge accumulation
Measurement	Production function	Normative-descriptive parameters

The basic assumption of the neo-classical economic growth theory was that each nation is able to overcome underdevelopment through reforms and the modernisation of its economic structures. Such an endogenous development process could be initiated by means of investing capital, which in turn was provided by an increased savings ratio, national redistribution processes, state credits, public expenditures or external inputs[71]. Growth, and hence development, are the result of the production function of the capital, labour and knowledge inputs [72].

According to MENZEL, two strategic approaches could be identified within the growth oriented development strategy; the approaches of balanced and unbalanced growth[73]. Both strategies focus on the entrepreneur, who needs incentives or must even be forced to invest in the industrialisation process. The unbalanced growth approach aims to overcome the 'natural' unwillingness to take risks and innovate. Induced incentives and competitiveness are seen as solutions to this. The balanced growth approach suggests creating incentives for entrepreneurs to invest by promoting the expansion of domestic markets for locally produced consumer goods, instead of focusing on the export of primary goods. The growth-oriented development strategy, and its various derivations, was successfully adopted in a number of

---

[67] The founder of this further development of the neo-classical growth theory was JOHN KEYNES. The paradigm resulting from it is known as 'Keynesianism'.

[68] This strategic approach has become known as the concept of 'growth first, redistribution later' and in a more political context 'industrialisation before democratisation' (MENZEL, 1993: 134).

[69] According to NOHLEN AND NUSCHELER (1993a: 64ff.)

[70] According to LACHMANN (2003: 184ff.)

[71] MENZEL (1993: 135ff.)

[72] In the neo-classical production function effective labour is the product of labour and knowledge.

[73] MENZEL (1993: 138f.)

countries during the 1950s and 1960s as far as economic growth is concerned and have been theoretically manifested in the "modernisation theories"[74].

In the 1970s first criticisms of this concept of growth and development arose from various sides[75]. Latin American scientists and development theorists declared that the development away from exporting primary goods towards import substitution through the production of domestic consumer goods could not be achieved with the proposed neo-classical strategies. Additionally, the democratisation process, which should theoretically follow increasing prosperity, did not take place[76]. The applied growth promoting measures and the successfully generated growth did not reach the smallholders, the poor or other marginalised groups and regions in these countries[77]. The macro-economists claimed that the only reasonable factor in the neo-classical input-output model for the explanation of income growth, namely effective labour, could not be explained and its behaviour was viewed as an externality[78].

Both, development theorists and macro-economists claimed that the modernisation theories provided insufficient explanations. The criticism expressed was directed at the growth theory itself and not at the idea of overcoming underdevelopment through economic growth[79]. The core statements of the critics were: 1) underdevelopment is not linked to the lack of capital; 2) the positive trickle-down effect of technological, economic and political development from advanced to less developed regions is not necessarily taking place; 3) international trade does not serve underdeveloped countries as an incentive for growth; and 4) the concept of catching-up through copying western models does not take into consideration the individual capacities and needs of the developing countries[80]. This criticism culminated in the statement that in reality the concept of development through growth turned into growth without development during the 1970s. A number of developing countries have improved their situation economically, yet the living conditions of the people, especially of marginalised groups, such as the poor and smallholders, have not improved.

---

[74] I.e. in various Latin American countries (MENZEL, 1993: 139). The GNP per capita grew by an average of 2.5% in Latin American countries in the 1960, while other regions of the world did not perform that well. In African countries the GNP per capita grew by 1.5% during the same time period and by 1.7% in South-Asian countries at (NUSCHELER, 1991: 19). In the 1980s the GNP growth per capita slowed down significantly in those same regions. Of 72 developing countries worldwide, only 21, mainly Asian countries, exhibited a growth per capita of more than 1% (MENZEL, 1993: 148f.)

[75] MENZEL (1993: 140ff.)

[76] BOECKH (1993: 118f.) and MENZEL (1993: 140)

[77] NUSCHELER (1991: 23)

[78] ROMER (2001: 98); According to KÖSTERS (1994: 117) about 50% of the national growth rate can be attributed to the progress of technology, which according to the neo-classical theory is integrated as an externality in the factor effective labour.

[79] NOHLEN AND NUSCHELER (1993a: 67)

[80] NUSCHELER (1991: 22)

On the basis of the above criticisms and the report by the World Bank's commission for international development in 1969[81], which criticised the common habit of the industrialised countries to link development aid and co-operation with international trade interests[82], a new basic understanding of development and underdevelopment emerged. The various concepts and approaches, which explain development in relation to multi-lateral dependencies and international factor exchange relationships[83], belong to the 'dependency' theory. This discourse on development changed the focus from internal factors of underdevelopment to external ones and widened the perspective of politicians and economists[84]. Major factors for underdevelopment according to the dependency theorists were the transfer of capital and resources from developing to industrialised countries, and the structural dependency of the developing countries' economies on the industrialised countries' demand and price policies[85].

Subsequent to these new development theoretical approaches, the development policies of the 1980s incorporated elements of both theoretical schools. The aim was to achieve a change in international trade and markets, and their effects on the developing countries. Important elements of the development strategy of the 1980s were: 1) to diminish the income gap between the north and the south; 2) to break the debt cycle, which hindered many countries in their development[86]; 3) to create an economic world order in which the developing countries control their own resources and participate in a liberalised market for agricultural goods and a regulated market for raw materials; 4) to increase self-reliance and south-south co-operation; and 5) to redistribute the economic added-value in developing societies[87]. The overall goal of the 1980s development strategy was to fight increasing poverty, which had resulted from the growth-oriented development strategies of the previous decades[88].

The outcome of the development policies of the 1980s was quite disappointing. At the end of the decade, the gap between north and south had increased, the overall quality of life of the people in developing countries had decreased and the number of people living below the poverty line with an income of less than US$ 1/day had increased. Refugees who tried to escape violence, poverty or ecological catastrophes were an additional problem[89]. Finally, the indebtedness of the developing countries increased at a time when world market prices for

---

[81] The report has become known as the 'Pearson-Report', named after the director of the commission for international development, LESTER PEARSON.

[82] NUSCHELER (1991: 21)

[83] The concept of factor exchange is based upon the trade theory and the two-region model, where regions, which are endowed with natural resources and labour export raw materials to regions with a high degree of manufacturing and capital capacity at a certain factor exchange rate (HARTMANN AND HÄRTER, 1989: 395f.).

[84] MENZEL (1993: 140)

[85] MENZEL (1993: 140)

[86] ...and which was a result of the neo-classical promotion strategy of capital transfer and lending in order to bridge the assumed capital gap.

[87] NUSCHELER (1993: 157ff.) and MENZEL (1993: 142ff.)

[88] MENZEL (1993: 152)

[89] In the 1980s the UNHCR counted an average of more than 15 million refugees per year (NUSCHELER, 1993: 160).

raw materials decreased[90]. In the late 1980s, when the aforementioned difficulties became apparent, international development policies reverted to neo-classical growth-oriented strategies[91]. This decision was confirmed by the performance of the East-Asian nations, which had developed rapidly and according to the structural patterns of the industrialised countries. This placed a question mark over the attitudes of the dependency theorists and their development strategies[92].

In the literature relating to development theories, the 1980s became known as the 'lost' decade[93]. This is due to the failure of development strategies and the underlying theories of development to effectively counteract underdevelopment and identify the causes of it, in spite of enormous political, financial and human efforts[94]. In the words of the development theorists: the great theories have failed to explain the development context[95]. The recognition that neither the 'dependency' nor the 'modernisation' theory could provide reasonable explanations and feasible strategies resulted in the rejection of the 'dependency' approach and the further development of the neo-classical growth theory into the 'New Growth Theory'.

The 'New Growth Theory' abandoned the idea of the perfect market and incorporated two important aspects, which were not taken into account by previous theories; the human capital[96] and technological progress[97]. Therefore, it pays tribute to the criticism of the development theorists and the recent debate[98]. The neo-classical growth theory regarded technological progress as an externality of the growth process, which is easily available to all entrepreneurs and all economic sectors[99]. The new theoretical approach on the other hand considers knowledge and technological progress as dependent factors. Knowledge and innovative capacity are direct and indirect parameters of development and growth at the same

---

[90] NUSCHELER (1993: 157ff.)

[91] MENZEL (1993: 152ff.)

[92] BOECKH (1993: 116ff.)

[93] NUSCHELER (1993: 156f.) and NUSCHELER (1991: 29ff.)

[94] MENZEL (1993: 152ff.)

[95] According to THIEL (1999: 21f.), any theory of development predicting and explaining human behaviour was condemned to failure as human behaviour cannot be predicted and takes place in a changing geographical , historical, political and economic context.

[96] The introduction of the factor 'human capital' into the growth theory was mainly driven by LUCAS (1988).

[97] The 'New Growth Theory' is based on an article on increasing returns and long-term growth by PAUL ROMER from 1986. In his approach ROMER focused on technological progress and its role in economic growth models (see also ROMER, 1990).

[98] The new factors were considered in the 'New Growth Theory', e.g. in research and development models (R&D-Models) to explain the neo-classical factor effective labour (knowledge*labour) and in the extension of the SOLOW-Model with the additional factors human capital and physical capital (ROMER, 2001: 98ff.).

[99] The basic production function of the original SOLOW-Model was (according to ROMER, 2001: 9):

Y=f (K(t); A(t); L(t)), with: Y=output; K=capital (including land); L=labour; A=effectiveness of labour (commonly known as knowledge).

Whereas the basic growth function of the 'New Growth Theory' was constituted of (according to LACHMANN, 2003: 16): Y=f (A(t); B(t); K(t); T(t)), with: Y= output; B= land; K=capital; T=technological progress. ROMER, 2001: 98ff.) described the various models of the 'New Growth Theory' and the different production functions resulting from them. However, the basic function is the reference and starting point for all of them.

time. Thus, the 'New Growth Theory' has overcome one basic discrepancy between growth and development theories.

The 'New Growth Theory' has provided the framework and basis for a number of mid-range theories and strategies[100], which focus on regional development approaches rather than macroeconomic explanation models[101]. From these mid-range approaches, the two new factors of the 'New Growth Theory' in particular, the human capital and technological progress, have been chosen as key research issues[102]. In *table 3* the basic differences between the neo-classical and the 'New Growth Theory' are listed.

Table 3. Major differences between the Neo-classical Growth Theory and the New Growth Theory[103]

	Neoclassical Growth Theory	New Growth Theory
Market	Perfect	Imperfect
Technological progress	Exogenous	Endogenous
Progress factors	Capital (including land), labour, (effectiveness of labour)	Capital (optionally divided into human capital and physical capital), labour, land, technology
Returns in scale	Constant and decreasing	Constant and not decreasing
Externalities	Negative	Positive
Growth	Increasing capital accumulation and capital transfer from rich to poor	Endogenous capital accumulation

The report by the 'Club of Rome' in 1973 on the 'Limits of Growth' and the negative relationship between economic growth and the world's environment and natural resources[104] initiated a process, which culminated in the United Nation's Conference on Environment and Development (UNCED) in Rio de Janeiro in 1992[105]. Since the UNCED, a more holistic approach, with an emphasis on development rather than on growth, has been accepted internationally. This 'Rio concept' of sustainable development[106] implies that only the combined achievement of economic growth, social justice and ecological sustainability will

[100] Mid-range theories attempt to explain a context or aspect under the consideration of a small, homogenous environment, which then makes the results comparable and transferable to similar conditions. It is based upon the results of research and case studies (ATTESLANDER, 1995: 47). For a detailed theoretical discussion of the understanding of mid-range theories see ESSER (2002: 128ff.).

[101] HAHN AND PREUß (1994: 71ff.); EßER ET AL. (1999); KAPPEL (1999)

[102] HEMMER AND WILHELM (1999: 251ff.)

[103] As compiled from KOSCHATZKY (2001: 64ff.) and ROMER (2001: 98ff.)

[104] For a detailed description of growth limiting factors as they were detected in the 1970s see LACHMANN (2003: 202ff.)

[105] MENZEL (1993: 141f.)

[106] The term 'sustainable development' was introduced by the Brundtland Report in 1987 (HARBORTH, 1993: 231)

ensure the future development of the world's nations[107]. The resultant discussion of sustainable development was quickly dominated by the globalisation debate, which focuses on the negative and positive effects globalisation may have on the individual countries of the world. The globalisation process, however, does not contradict the concept of sustainable development. Rather, the concept of sustainable development is an approach to guide the globalisation processes[108].

One consequence of the development decade[109] of the 1980s was the adoption of enhanced neo-liberal[110] adjustment strategies in the 1990s[111], which constitute a characterising element of the globalisation process as it is known today[112]. The neo-liberal orientation of this development strategy, as it was also promoted by the World Bank until the late 1990s, is characterised by: 1) the rule of the market, 2) increased deregulation, 3) increased privatisation, 4) decreased public expenditure for social services, and 5) a general increase in individual responsibility[113]. This strategic approach to promote development through economic growth has threatened the sustainability of social justice and the ecological environment in many countries of the world[114]. Thus, the World Bank changed its position in the late 1990s, away from the neo-liberal approach to a more social and integrated perspective, and strategic orientation[115].

In 1993, NOHLEN AND NUSCHELER gave a definition of development integrating the experiences of the development decades from the 1950s to the 1980s. They defined development as: "*... the endogenous evolvement of the productive forces to provide the whole society with indispensable material and cultural goods within the framework of a social and political order, which guarantees equality of possibilities to all members of this society and*

[107] BOHNET (1999: 97) and HARBORTH (1993: 231ff.)

[108] As explained by SOLIMANO (1999)

[109] International development and development co-operation policies began in the 1960s, which is known as the first development decade, followed by the second development decade in the 1970s, the third in the 1980s and the fourth in the 1990s (MENZEL, 1993: 131ff.).

[110] ADAM SMITH first established the concept of economic 'liberalism' in 1776 ('*The wealth of nations*'). Its core statement was: no restrictions on manufacturing, no barriers to commerce and no tariffs. According to the liberal idea, free trade was the best way for a nation's economy to develop.

[111] MENZEL (1993: 154f.)

[112] BOHNET (1999: 104ff.) defines 'Globalisation' as follows (translated from German): "Globalisation is the process of increasing worldwide economic and non-economic integration. This process is taking place with increasing velocity." BOHNET distinguished between four types of globalisation: 1) cultural, 2) political, 3) ecological, and 4) economic. He emphasised that all four types are closely interlinked.

[113] The neo-liberal development approach has become a substantial part of the World Bank's internationally applied economic policies. The World Bank's so-called 'Washington Consensus' postulated the adoption of neo-liberal development strategies by developing countries and formed the platform of the individual adjustment policies of these countries (UNMÜSSIG AND WALTHER, 1999: 247).

[114] As described according to social aspects for Latin America by BOECKH (1999: 93ff.), and discussed under ecological and social aspects by proponents of 'political ecology', such as KRINGS AND MÜLLER (2001).

[115] Since 1998 the World Bank has revised its position and promotion of the neo-liberal adjustment policies. This new trend has become known as the Post-Washington Consensus and emphasises cultural societal and institutional aspects, in addition to the economic development (STATZ, 2000: 40f.).

lets them participate in the political decision-making and contribute to the creation of the common prosperity.'[116]. This understanding of development is used as a reference for this research.

There is empirical proof that economic growth is an important factor of development. The World Bank's development report of the year 2000 entitled 'Attacking Poverty' shows the positive correlations between economic growth and development indicators, such as the reduction of poverty, health care indicators, enrolment rates of children in schools, etc. The most important correlation is that of growth and the reduction of poverty[117]. These correlations, however, provide only a partial explanation of the factors influencing development. In various cases, economic growth did not result in improved development indicators, and increasing per capita income does not automatically mean that the income is equally distributed within society[118]. In general, the results of the World Bank's experiences and investigations reveal that development is almost impossible without economic growth, but economic growth alone does not guarantee development.

2.1 Regional Development and Concepts of Competitive Advantages

The development theories have failed partly because they depended too much on the economic growth theories[119]. Hence, new approaches like the concept of 'sustainable development' have achieved increasing recognition. The new development theoretical approaches of the early 1990s aimed at explaining and analysing development focused on the three aspects of 'sustainable development': ecology, society and economy. These approaches were based on the early concepts of rural development approaches, which focused on the strengthening of the agricultural sectors in order to increase productivity and contribute to the national economy; the 'agricultural development' approach originated in the 1950s and 60s[120]. This approach was extended during the 1970s by integrating socio-political and socio-economic aspects into the support strategies. The concept of regional development as it is known today emerged from that[121].

A special focus of the concept of integrated regional development is the optimal employment of locally available resources and the equal distribution of the benefits derived from them. The primary aim of this concept is to cover the basic needs of the local population and to reduce poverty[122]. According to economic considerations, the optimal employment of the

[116] NOHLEN AND NUSCHELER (1993a: 73)

[117] THE WORLD BANK (2001: 45ff.)

[118] THE WORLD BANK (2001: 49ff.)

[119] HAHN AND PREUß (1994: 71)

[120] HAHN AND PREUß (1994: 77)

[121] HAHN AND PREUß (1994: 78)

[122] HAHN AND PREUß (1994: 79f.)

resources available locally and the development of human capital represent factors, which increase the competitive advantage of a region on a national, as well as international scale. This is closely related to the competitiveness of the local economic sectors.

Subsequent to the debate of the development theories in the 1980s, the growth and the development theories have converged. This has happened through the application of mid-range perspectives in the development theories and through the broader perspective of the 'New Growth Theory', in respect of the additional factors for growth, namely knowledge and technology. Both propose fostering the implementation of the necessary institutional framework and strengthening the role of the local population. Nevertheless, the primary aim of the development strategies adopted by international organisations, development co-operations, as well as national and international development policies since the 1990s continues to be economic growth[123].

The gap between industrialised and developing countries, formerly described as a capital gap, is now understood to be a gap in knowledge and technology[124]. However, this understanding is already a further development of the basic assumptions of the 'New Growth Theory'. The 'New Growth Theory' itself only incorporated the new factors through the application of the old instrument of capital-oriented production functions. Hence, the causes of innovation, technological progress and human capital development remained unexplained in the 'New Growth Theory'[125].

Closely linked to the emergence of the 'New Growth Theory' is the concept of competitive advantages by PORTER, which achieved increasing recognition in the 1990s. This approach is based on the industrial cluster concept, which focuses on the performance of the industries and firms of a region, rather than on the performance of the national economy[126]. The cluster concept is described in *chapter 2.2*. Important for the understanding of this research is PORTER'S structure of the determinants of competitiveness to facilitate the identification of differences in the economic success across countries. The definition of the economic success of an industry or firm, according to PORTER, is the ability to maintain long-term profitability and a high market share, accompanied by continuous innovation and cluster upgrading[127]. This upgrading process increases the competitive advantage of the sector and can be divided into 1) factor-driven competitive advantage, which refers to the availability of resources, such as labour, raw materials, etc.: 2) investment-driven competitive advantage, which refers to the monetary and the non-monetary infrastructure; and 3) innovation-driven[128] competitive

[123] See THE WORLD BANK (2001: 31ff.) and NUSCHELER (1993: 169ff.)

[124] FERRANTI ET AL. (2003: 1ff.)

[125] For a more detailed critique of the 'New Growth Theory' see KOSCHATZKY (2001: 76ff.)

[126] PORTER (1990: 148ff.)

[127] PORTER (1990: 278)

[128] For a detailed discussion of the term innovation and the driving factors behind the innovation process see *chapter* 3.2.

advantage, which refers, e.g. to the state of R&D in a sector, enterprise network structures, etc. These three stages of upgrading follow one another in time[129].

The concept of competitive advantage sets the conditions of economic sectors of different regions in relation to each other. The core statement of the concept is that the competitiveness of a nation is based on the competitiveness of its economic sectors[130]. The analysis of the conditions of these sectors allows their comparison and the initiation of measures to change, usually to improve, any of these conditions in order to achieve advantages regarding the two success factors profitability and market share. The Diamond-Model by PORTER describes the four major determinants of competitive advantage of a sector's firms as[131]: 1) the factor conditions, in terms of basic factor availability, such as natural resources, population or climate, and advanced factors, such as communication infrastructure, research facilities, etc.; 2) international and domestic demand conditions; the domestic demand is the main motor of innovation; 3) the availability of related and supporting industries, which in turn are internationally and nationally competitive; and 4) the firm's own strategy, structure and competitive behaviour within the cluster. Additionally, important outside forces that form the firm's environment are[132]: the government and its policies, chance and international business activities.

The competitive advantages have been included in nearly all of the development strategies aimed at enhancing regional economic growth[133]. Social or ecological elements have been added to this. The concepts of industrial/economic clusters and PORTER'S 'competitive advantages' are the bases for one of the most recent approaches, the concept of 'systemic competitiveness'[134]. Based on empirical studies of economic clusters, EßER ET AL. developed this concept[135]. The development of human capital and the ability of the firms to develop new products, processes and organisational elements are the focus of this approach. The role of the individual actors and their interactions are seen as the key factors for upgrading competitive advantages: '*interactions create competitiveness*'[136]. This statement is a result of the deficits of the great theories, including the 'New Growth Theory' and the 'competitive advantages' approach[137], to explain the role of the processes underlying the creation of innovation,

---

[129] PORTER (1990: 545ff.)

[130] PORTER (1990 : 33)

[131] PORTER (1990: 71ff.)

[132] PORTER (1990: 126ff.)

[133] GTZ (2001a: 14)

[134] EßER ET AL. (1999: 149f.)

[135] EßER ET AL., 1992; HILLEBRAND, 1991; MESSNER, 1990 and MEYER-STAMER, 1991

[136] EßER ET AL. (1999: 152ff.)

[137] With the exception of the government and the entrepreneurs, other societal actors, such as associations, universities and other organisations are not included in PORTER'S approach (GTZ, 2001a: 16).

competitiveness and economic growth. Increasing competitiveness is seen as a societal process, rather than a mere economic one[138].

The major characteristics of the 'systemic competitiveness' approach are[139]: the analytical differentiation of four levels (micro, meso, macro and meta-level[140]), the consideration of regional and global influences on the analysed sector, and the importance of the interactions within the sector as a major driving force behind competitiveness. In *table 4*, the major features of the approaches 'competitive advantage' and 'systemic competitiveness' are compared.

Table 4. Features of 'competitive advantages[141]' and 'systemic competitiveness[142]'

Feature	'Competitive advantages'	'Systemic competitiveness'
Basic economic understanding	Clusters/Industrial districts	Economic clusters
Considered actors	Enterprises, government	Micro, meso, macro-level actors
Parameters of competitiveness	Long-term profitability and market share of a firm	The system's sustainability and welfare
Source of competitiveness	Driven by investment, innovation	Driven by interaction
Analytical levels	Factor and demand conditions, supporting industries and government policies	Economic and political institutional considerations of the micro, meso, macro, meta-level
Analytical approach	Micro and macro economic data	Actor and interaction centred research
Strategy	Growth-oriented	Development oriented

The assumption that actor interactions are the driving force behind competitiveness and economic growth is based on economic research, which emphasises the patterns of clustering industries, and the behaviour of the firms in these agglomerations. This behaviour may vary between the two extremes rivalry and partnership. In the economic sciences research into industrial districts and clusters studied the relationships between firms and their environment, and how these interrelations influence their own economic performance and that of the economic sector as a whole.

---

138 A consideration closely related to the ideas of innovation economics and evolutionary economics. See *chapter* 3.2.2.

139 According to EßER ET AL (1999: 152ff.) and GTZ (2001a: 27ff.)

140 For a detailed explanation of these levels see *chapter* 3.1.

141 Based on PORTER (1990)

142 Based on EßER ET AL. (1999)

2.2 Interactions in economic Clusters

As mentioned in *chapter* 2.1, the concept of economic clusters as an explanatory model for the regional agglomeration of firms in economic sectors is the basic concept, which serves as the basis of the 'competitive advantages' and 'systemic competitiveness' approaches. Although there is no common definition of the term cluster in the literature[143], the cluster concept is the basic concept of a number of explanatory models dealing with regional economic growth and development. Basically, a cluster describes a *'local concentration of certain economic activities'*[144]. Hence, cluster is primarily a spatial definition. The proximity of the firms frequently attracts policy makers who choose these agglomerations as promising objects for economic promotion programmes since transaction costs[145] are low and the possibility for joint action is high. Beyond the context of the spatial definition, the basic advantage of the agglomerating firms is seen in the interactions among the firms and the positive external effects emanating from the existence of a pool of common knowledge or the common human capital pool[146]. A definition proposed by ALTENBURG AND MEYER-STAMER describes a cluster as: "... *as sizeable agglomeration of firms in a spatially delimited area, which has a distinctive specialization profile and in which inter-firm specialization and trade is substantial*"[147]. This is the definition of the term cluster employed in this study. In order to avoid confusion with other concepts of regional economic agglomerations, *table 5* provides an overview of the most common concepts and their basic differences.

[143] GTZ (2001a: 22) and ALTENBURG AND MEYER-STAMER (1999: 1694)

[144] ALTENBURG AND MEYER-STAMER (1999: 1694)

[145] The term 'transaction costs', which first was introduced by RONALD COASE (1937), is used in the 'New Institutional Economics' sense, as established by DOUGLAS NORTH (1988). The 'New Institutional Economics' constitute an important structuring element of the cluster concept and the various concepts of competitiveness as described in *chapter* 2.2 (GTZ, 2001a: 13 and EßER ET AL., 1999: 149). Furthermore, it has found its application in development studies and research (NORTH, 1993)

[146] ALTENBURG AND MEYER-STAMER (1999: 1694)

[147] ALTENBURG AND MEYER-STAMER (1999: 1694)

Table 5. Concepts of regional economic agglomerations[148]

Feature	'Industrial districts'[149]	'Clusters'	'Regional innovation systems'[150]
Economic parameters	Neo-classical economy	Market shares and collective profitability; upgrading levels	Collective innovation process and technological advantages
Reason for competitiveness	High degree of specialisation and co-operation, flexibility	Interactions among specialised firms (usually SMEs) in form of competition and co-operation	Value chain of SMEs and large scale firms and plants with close co-operation networks
Basic pre-conditions	Geographical agglomeration, trust, local support-systems[151], tradition of manufacturing and entrepreneurship	Geographical agglomeration of firms of an economic sector	Network of firms in a sector and its support system, trust, pool of educated human capital
Empirical experiences	Industrialised countries: i.e. in Europe: Italy	Several examples in industrialised and developing countries	Examples from industrialised countries
Transferability of the concept[152]	Limited: only few examples worldwide, complex pre-conditions required	High applicability due to relatively few economic requirements and focus on competitiveness	Limited: special pre-conditions and requirements needed and difficult quantification of the results

As shown in *table 5*, interactions either in form of loose co-operations or close network structures are key factors of successful regional agglomerations of firms. Consequently, this aspect is increasingly included in economic promotion strategies. The concepts of regional economic agglomerations presented provide the basic information for policy makers to intervene in the development processes of existing firm agglomerations and promote a certain direction of development. The common aims of such interventions are:

- To improve and adjust the administrative and institutional conditions as are usually achieved through national programmes in co-operation with international development organisations. These approaches focus on the socio-economic aspects of business development[153], and the entities regulating international industry and

[148] According to the chapter on local economic sector promotion in GTZ (2001a: 13ff.)

[149] See KOSCHATZKY (2001: 186ff.). Recommended literature on industrial districts includes: PYKE, BECATTINI AND SENGENBERGER (1990); BECATTINI (1990); BRUSCO (1990); COOKE AND MORGAN (1994); and BIANCHI (1998). (GTZ, 2001a: 21f.)

[150] For a detailed discussion of regional innovation systems see *chapter* 3.2.3.

[151] Local support systems are made up of educational facilities and universities, which provide human capital, know-how and R&D facilities to the regional firms (GTZ, 2001a: 19).

[152] Modified and expanded according to KOSCHATZKY (2001: 207)

[153] See GTZ (2001a: 54)

commerce. The WTO and the IMF are responsible for the implementation of such national adjustment policies with liberal economic orientation[154].

- To directly promote the entrepreneurs and their interactions, especially as carried out by the UNIDO and ILO[155]. They aim for competitive advantages based on networks and the business location[156]. In particular, these promotional approaches take into account the economic sustainability and self-sufficiency of the promoted entrepreneurs, which are generally micro, small and medium sized enterprises[157].The cluster concept is rooted in the discussion of the 'industrial districts', which was originally developed to explain the success of agglomerations of small and medium sized firms in northern Italy[158]. Most of the research conducted on the cluster concept so far has been done in European countries and regions. In the second half of the 1990s the concept was increasingly applied to analyse local agglomerations of firms in developing countries[159]. However, the experiences originally made in the industrialised countries were not transferable to developing countries[160].

Other explanatory and promotional models of economic sector agglomerations (as e.g. the 'industrial districts' and the 'regional innovation systems') were even less suited to development strategies, because these models require advanced and elaborate support-systems, such as universities and R&D facilities, and a deeply rooted entrepreneurial and manufacturing culture. Furthermore, the presence of trust[161] and a culture of co-operation are basic pre-conditions for the adoption of these models. Since the cluster approach seems to be quite indifferent towards the existence or non-existence of a sophisticated supporting infrastructure and the integration of large-scale entrepreneurs, the cluster concept has been applied in several cases to explain the performance and development of economic sectors in industrialised, as well as in developing countries[162].

In Europe, the cluster concept is usually applied to explain the economic performance of regional industrial branches. VERBECK, for example, compared clusters in several European countries. Their innovative performance was analysed, assuming that innovation is the

---

[154] See DRESCHER (2002) and HAUSER (2002:15)

[155] GTZ (2001a: 54)

[156] For examples of such activities see: KAPLINSKY AND READMAN (2001); HUMPHREY AND SCHMITZ (2000); CEGLI AND DINI (1999) or LEVITSKY (1993)

[157] GTZ (2001a: 55)

[158] GTZ (2001a: 17)

[159] GTZ (2001a: 23)

[160] GTZ (2001a: 24)

[161] The issue of trust and co-operation is discussed in detail in *chapter 3.1* in the context of the meta-level in the concept of systemic competitiveness.

[162] For German cluster policies see the publications by BRENNER. For a compilation of cluster policies and strategies in OECD countries see VERBEEK (1999) and ROELANDT, GILSING AND SINDEREN (2000).

driving force behind the economic development of clusters[163]. He found that there is no significant correlation between the existence of sophisticated network structures, i.e. the supporting systems and the innovative performance of the investigated clusters. VERBECK explained this structure-performance-paradox with the existence of non-measurable factors for innovative behaviour, such as the general economic atmosphere in a region or nation, the small sample size he selected and, finally, the possibility that innovative behaviour in general is much more resistant to the support-system than it has been assumed.

The emphasis of explaining cluster development is then on actor behaviour and the endogenous cluster developments, which determine economic and technological development. Knowledge and its distribution within the cluster, independent of promotional measures, are an important aspect in this context. Furthermore, the degree of the internal structural organisation of a cluster is of importance. BRENNER investigated economic clusters in Germany and concluded that cluster development follows an endogenous evolutionary path rather than reacting to external stimuli and policies[164]. This interpretation is based on the theory of 'evolutionary economics', which is discussed in *chapter 3.2*. BRENNER'S simple conclusion is: "*If policy measures are launched too early or too late, they turn out to be clear waste of money*"[165].

The experiences made in Europe have led to the awareness that policy makers and promotional programmes usually focus on external factors, such as the main customers, demographic trends, the products and services markets and the internal cluster factors, like the key resources (hard factors). Nevertheless, it is the societal factors, namely the social and human capital, knowledge and the distribution of knowledge, which are responsible for economic performance and development[166].

In contrast to the European experiences, the production process of clusters in developing countries constitute only a small part of the added-value. There are only few innovations with regard to the products and processes. The degree of specialisation of the individual firms is low and co-operation between the firms is characterised by mistrust[167]. HUMPHREY AND SCHMITZ concluded for clusters in developing countries that increased co-operation between firms improved their performance and competitiveness[168]. Based on their experience of clusters in developing countries, ALTENBURG AND MEYER-STAMER developed a classification of clusters typically found in Latin American countries[169]:

---

163 VERBECK (1999: 49f.)

164 BRENNER (2000; 2002; 2003) and BRENNER AND GREIF (2003)

165 BRENNER (2003: 343)

166 See MOROSINI (2003)

167 GTZ (2001a: 23)

168 HUMPHREY AND SCHMITZ (2000: 8f.) base their conclusion on four cluster studies carried out in India (KNORRINGA, 1999), in Pakistan (NADVI, 1999), in Mexico (RABELLOTTI, 1999) and in Brazil (SCHMITZ, 1999).

169 ALTENBURG AND MEYER-STAMER (1999: 1695ff.)

Survival clusters of micro and small enterprises, which produce low-quality consumer goods for local markets. These clusters are characterised by high numbers of firm entries and exits. In general, firms are organised informally. The degree of specialisation and inter-firm co-operation is low. These clusters are the most common in developing countries and their size may vary from a few hundred to several thousand small firms.

Advanced and differentiated clusters of mass producers are characterised by the broad range of members, from micro enterprises to large-scale 'Fordist[170]' industries. They mainly produce for the domestic market and benefit from import substitutions. Market liberalisation is a threat to their international competitiveness. The innovation rate, as well as the degree of interaction and co-operation is low.

Clusters of trans-national corporations are usually world-class manufacturers and serve national and international markets. They account for high added values and global integration in value chains. However, there are few links to small and medium sized local enterprises. In these clusters, large factories or plants are the dominant actors. The barriers to entry to such clusters are comparatively high.

In relation to the focus of this research, the concept of survival clusters will contribute significantly to the reference system of the research concept. MEYER-STAMER conducted a number of studies in all three types of clusters in Germany and Latin America[171]. He stated that for the successful promotion of economic sectors in developing countries, the cluster is the most suitable analytical tool. He made the following recommendations for promoting clusters[172]:

- A shift from subsidy and incentive-oriented promotion instruments to a mix of actor, institution and incentive-oriented instruments.
- Making use of local investment potential instead of attracting external investors.
- The improvement of the business location factors in selected regional centres instead of scattered projects. This includes the traditional hard business location factors, as well as the supporting systems and soft factors.
- The exchange of experience between these regional centres.

---

[170] The term 'Fordist' is used to denote the highest possible (and profitable) specialisation of labour within the production process (MEYER-STAMER, 1999a: 7f.).

[171] ALTENBURG, HILLEBRAND AND MEYER-STAMER (1998); ALTENBURG AND MEYER-STAMER (1999); MEYER-STAMER ET AL. (1996); MEYER-STAMER (1996a; 1996b)

[172] MEYER-STAMER (1999b: 21f.)

The aforementioned measures are aimed at the promotion of the parameters relevant to the economic success of the cluster: market share, collective profitability and upgrading, which are indicators for the cluster's economic performance and competitiveness. Collective profitability and market shares are much easier to assess and measure than 'upgrading'. Upgrading comprises the change of a firm's economic activities from lower to higher efficiency or from one type of activity to another. HUMPHREY AND SCHMITZ defined three types of cluster upgrading[173]:

- Process upgrading: The adoption of new technologies or the reorganisation of the production process
- Product upgrading: The production of new or more sophisticated products
- Functional upgrading: The firm itself changes its position in the cluster, e.g. from a provider of raw materials to a producer of semi-processed goods

The collective upgrading of firms leads to cluster upgrading. Such an upgrading process may result, for example in a shift from survival to industrial clusters, or the cluster remains a survival cluster, but the market share of its production is increased. In many cases the long-term aim of upgrading is the improved international competitiveness of the cluster and its integration into international value-chains[174]. This assumption is based on the experience that integration into global value chains and simultaneous improvement of the SMEs efficiency results in income growth[175].

The task of upgrading a cluster must take into consideration the cluster as a system of classical production factors, human capital and a common technological pool. The whole system depends on its actors and their upgrading capacity, which is linked to their access to knowledge and capital sources. According to the cluster concept, this is facilitated by interaction in the form of competition and co-operation. The employment of the knowledge and the capital within the enterprises enables them to upgrade their products and production processes. Thus, the process of upgrading is closely related to the innovation processes in a cluster[176]. The results of upgrading processes can be measured or at least described through the indicators provided by innovation research and economics, amongst others.

---

[173] According to HUMPHREY AND SCHMITZ (2000: 3f.)

[174] HUMPHREY AND SCHMITZ (2000: 6ff.)

[175] KAPLINSKY AND READMAN (2001: 37)

[176] KAPLINSKY AND READMAN (2001: 27)

2.3 The Role of Forests in Development and Growth

Current development strategies of the World Bank, the IMF and national governments are based on the 'New Growth Theory' and the concept of sustainable regional development. Elements of centre-periphery approaches and the new trade theory constitute essential parts of these strategies. This understanding of 'development through economic growth' reflects the convergence of the neo-classical growth theory and the 'dependency' theories. The assumption of both, that development strongly depends on exogenous factors, has been replaced by a basic understanding of endogenous development processes[177].

In addition to the traditional production factors other core elements of the 'New Growth Theory', the human capital and technological progress, have been examined in case studies and research in industrialised countries[178]. In developing countries comparable research on innovation and entrepreneurship has mainly been conducted in metal, textile and agro-industrial sectors[179]. Only little research into sectors producing and manufacturing natural raw materials as for example forest resources has been conducted so far[180].

Forest resources play an important role in economic and societal development as they can be exhausted temporarily and repeatedly regenerated[181]. Before the 1970s and the report by the 'Club of Rome', international discussions on the economic importance of forest resources were not common. However, decreasing forest resources and the negative economic impacts related to the decline of forest resources have been documented since ancient times[182]. On the other hand, forests have increasingly been perceived as an obstacle to cultural and economic development, i.e. in the context of rapid discovery and settlement processes, such as in North and South America during the last five centuries[183]. A full examination of such a time frame

---

[177] In the neo-classical theory growth was considered endogenous. However, as underdevelopment was the result of the lack of endogenous potential, i.e. the supply and demand for capital, the development strategies derived from it focused on external inputs into the developing countries.

[178] VERBECK (1999); Brenner (2000); BRENNER (2002); Brenner (2003); BRENNER AND GREIF (2003); MEYER-STAMER (1999a and b); etc.

[179] ALTENBURG AND MEYER-STAMER (1999); BUITELAAR (2002); BORTOGARAY AND TIFFIN (2000); MEYER-STAMER (2002); MEYER-STAMER AND SEIDEL (2002); GUIAPATIN (2002); CEGLI AND DINI (1999); HUMPHREY AND SCHMITZ (2000); KNORRINGA (1999); NADVI (1999); RABELLOTTI (1999); SCHMITZ (1999); MOROSINI (2003); etc.

[180] Exceptions are BONITA ET AL. (2002) and KOECHLIN (2000)

[181] ENDRES AND QUERNER (1993: Vff.)

[182] For example, the importance of forest resources in the construction of merchant and war navies in ancient Greece and the Roman Empire, but also in Spain, France, the Netherlands, Great Britain and other countries between the 15th and 18th centuries resulting in deforested and devastated landscapes around the Mediterranean Sea. The forest resources needed to provide energy for high value-adding industries, such as the glassblowing and smelting industries in Germany of the 18th and 19th century, led to a severe shortage of wood. Finally, the examples of the great number of tropical forest fruit and other useful plants - their domestication and large-scale cultivation resulted in the destruction of large areas of natural forest – a process, which is still going on. See SCHMIDT (2002: 10ff.) and WILLIAMS (2003).

[183] WILLIAMS (2003: 210ff. and 371ff.)

would exceed the boundaries of this research. Therefore, this chapter concentrates on the period from the 1950s to the 1990s, to complement the theoretical discussion of the previous chapters.

Traditionally, the exploitation of raw materials, such as minerals, timber or foodstuffs, was the only contribution of natural resources to the development of a region or a nation. At the same time, this exploitation was usually a first step towards the transformation of "wilderness" into settlements and cultivated land. According to neo-classical growth theory, the most important resource forests provide in economic terms is the land itself. This land, which represents one of the traditional production factors, has to be used efficiently to optimise the production function. The opportunity costs[184] of any utilisation option have to be calculated in advance. Consequently, the forest resources, which provided economically necessary and indispensable raw materials, were subject to optimisation calculations, aimed at the maximisation of the returns per land unit.

This implies that the less forest raw materials are valued by an economy or society, the higher is the possibility that the forests will be cleared for other purposes with lower opportunity costs and higher returns per unit of land. The classic and still common economic decision-making tool with regard to forest resources and their management is the 'FAUSTMANN Formula[185]', which is based on calculations according to the land rent theories[186] and aims at ensuring sustainable[187]' economic returns from the forest resource base. In industrialised and developing countries more sophisticated or modified versions of the 'FAUSTMANN Formula' continue to be an important economic decision-making tool [188], such as the definition of rotation periods, as well as individual tree and stand volume maximisation[189].

Whereas the deforestation processes during the period from the ancient to the pre-industrial age took place over centuries and were restricted to relatively small areas, the deforestation processes of the 20th century have destroyed 20% of the world's forest cover and more than

---

[184] Opportunity costs describe the profits from a production alternative not engaged in, as compared to those of the actual option (either at micro or macro-economic level) (GABLER WIRTSCHAFTSLEXIKON, 1993: 2495).

[185] In 1849 MARTIN FAUSTMANN developed a formula to calculate the optimal rotation period under consideration of the expected profitability of the production factor land. As a consequence the net present value and the asset value of forest resources played an important role in forest management and land use decisions. For a detailed discussion of the formula and its implications for ancient and modern forestry see NAVARRO (2003).

[186] NAVARRO (2003: 51ff.). The land rent theory itself was derived by RICARDO (1817), a representative of the classical economic theory according to ADAM SMITH (1776). Hence, in trade theory, RICARDO goods are the primary goods produced through the intensive employment of the production factor land.

[187] The term 'sustainable' in this context is based on German forestry of the 19th century and one of its first theorists CARLOWITZ (1713), who understood sustainability as the long-term provision of forest raw materials - an understanding, which is quite different to the 'sustainable development' paradigm of the 1990s. See SCHMIDT (2002: 6ff.).

[188] As modified by OHLIN in 1921, for example. The original FAUSTMANN formula without intermediate treatment of the stand was: Lo=Rh-Ca*(1+i)/(1+i)-1, with: Lo=expected value of the forest land without trees; Rh=the revenue of the timber harvest; Ca=afforestion costs at the end of the rotation; (1+i)=discrete interest factor in dependency from T; T=rotation period. (NAVARRO, 2003: 38f.)

[189] NAVARRO (2003: 36ff.)

50% of the tropical rainforests in almost 50 years[190]. The 1980s was the decade, which saw the highest ever deforestation rates in the world's developing countries[191]. This fact can be related to the failed development policies of that decade and the return to neo-liberal economic growth policies at the end of the decade[192]. The deforestation processes of the last decades led to large-scale transformations of forest into agricultural land, as well as degraded and scattered forest resources[193].

According to the concepts of competitiveness and growth, development strategies have to consider the optimal utilisation of local resources to gain a comparative advantage[194]. In Europe, forest resources have partly lost their economic importance for the national economies and for the livelihood of the majority of the populations. However, forest use and forest-related sectors are integral to regional development strategies, and contribute to regional economies, as illustrated in case studies by the European Forest Research Institute (EFI)[195].

In developing countries a number of strategies have been applied to ensure the survival of the forests, which range from pure protection areas and biosphere reserves with buffer zones to integrated community forest management approaches[196]. The management strategies aim towards the sustainable use of the resources, as well as the creation of employment and income generation. The socio-economic aspect of the equal distribution and the generation of added value are emphasised. The utilisation of forest resources as an integral part of regional and national development strategies pursues two major aims[197]: 1) the promotion of the sustainable use of the resources; and 2) the promotion of the local economic sectors based on forest resources. STATZ summarised the most important arguments in favour of this strategy[198]:

- Incomes and returns from forestry and the related manufacturing and service sectors contribute to the national economy. The higher this contribution is, the higher the valuation of the resources should be, and consequently the emphasis on long-term sustainable use and protection.

---

190 See WILLIAMS (2001).

191 See figures in FAO (1999 and 2001)

192 I.e. in the Latin American context (Statz, 2000: 59).

193 Examples for such areas are the 'parklands' in Sub-Saharan Africa, agroforestry housegarden systems in South-East Asia or the bamboo resources in the Colombian coffee growing region and many more rural areas with integrated agricultural and forestry production systems.

194 ANDERSEN (1996a: 5ff.)

195 HYTTINEN, P., MONONEN, A. AND PELLI, P. (1996)

196 For the most recent experiences with these types forest management approaches integrated in development strategies, see THE WORLD BANK (2001: 91ff.); DOLOM (2003: 22ff.); LUJAN ET AL. (2003: 30ff.); AKUMSI (2003: 37ff.); HEADLEY (2003: 44ff.); LAI (2003: 51ff.); FERROUKHI (2003)

197 STATZ (2000: 36f. and 59)

198 STATZ (2000: 14f.) is focused on the use of non-timber forest products (NTFPs). For the purposes of this study only the arguments related to all forest products were adopted.

- The harvesting, processing and marketing of forest products are labour intensive and, therefore, provide jobs and income for a large number of people, which is of high economic and social importance, especially in structurally weak rural areas of developing countries.
- Regular small-scale rural forestry (as practiced in the degraded and scattered forests today) contributes to the rural livelihoods and is easily integrated into rural farming systems.
- Forests as an income source offer economically and socially marginalised groups the opportunity to improve their situation. This refers especially to the use of non-timber products, whereas the use of timber is usually limited to privileged actors.
- The promotion of regional forestry and forest-based economies offers the possibility for grass root level participation in institutions-building and decision-making, which is crucial for the strengthening of social capital and society as a whole.

Value-adding in the wood processing industries is linked to advanced technology. In contrast to industrialised countries, in developing countries these technological facilities are not very widespread and are difficult to acquire. In order to examine the role of forest resources as a basis for economic and industrial development the FAO compared the production of round wood with the production of processed wood products in OECD and non-OECD countries. The result was that non-OECD countries have export surpluses of round wood, whereas OECD countries usually import additional round wood to meet the demands of their wood processing industries[199]; a sign of the higher comparative advantage of these countries as business locations for value-adding processes in the wood processing sector. In accordance with the classical and neo-classical growth theories, raw forest materials are traditional goods as classified by RICARDO, which have to be exchanged for high value products through trade. According to the 'New Growth Theory', these raw materials represent the local factors necessary to meet the local population's demand, on the one hand, and to participate in national and international trade on the other[200].

To overcome this discrepancy the adopted development promoting strategies consider a variety of measures, of which some are based on neo-classical growth promotion strategies (through external input of capital and technology), whereas others focus on education and extension of knowledge[201]. In combination, these strategies are a transfer of technology and knowledge. The strategies for forest sector promotion in developing countries are built on the experiences made in the forest sectors of the Scandinavia and other industrialised countries.

[199] WHITEMAN, BROWN AND BULL (1999: Chapter 3.2.3)

[200] STATZ (2000: 39); KAPPEL (1999: 281ff.) saw in the increased production of Ricardo goods an economically feasible option to promote growth at the beginning o f the new millennium.

[201] See FERRANTI ET AL. (2002: 49ff.) and KÄRKKÄINEN (1996). For measures to overcome technology and knowledge gaps see FERRANTI ET AL. (2003)

In modified form, these experiences are considered to be suitable models for further development of other forest sectors also[202].

In PORTER'S classification, natural resources play an important role with the regard to the performance and competitiveness of industrial sectors. He calls these factors the inherited factors of a region. The forest products sector is an upstream industry according to the cluster classification chart by PORTER. In this chart the forest products sector is vertically linked to other industry sectors producing and processing primary goods, such as metal and petroleum. It is horizontally linked to industrial sectors with supporting functions, such as power generation and transport[203].

BONITA ET AL. (2002) investigated the forest sectors in five Latin American countries, applying the analytical approach of PORTER'S competition model based on clusters. Based on the Scandinavian experience, the following key elements of development and competitiveness in the Latin American forest sectors have been identified[204]: 1) sound macro-economic policies for business development; 2) competitive pricing of forest products and services in domestic and international markets; 3) close co-operation with suppliers of required specialised inputs; 4) a functioning network of related supporting industries; and 5) co-operation amongst the actors in a cluster. Again, co-operation and interaction are crucial elements of successful development and growth. Various obstacles and problems hinder the implementation of such development measures in the Latin American forest sector. *Table 6* lists the most important problems experienced by five Latin American countries.

In summary, *table 6* shows that the main problems of the forest sectors in these five Latin American countries are the lack of co-operation amongst firms, the general neglect of the forest resources as a source of employment and income generation, as well as a low technological level (including the management of human capital). Altogether, there is a general lack of a value-adding capacity at the micro-level and, hence, a lack of competitiveness. At the institutional level, the major deficits are the lack of or inappropriate forest management legislation and poor land tenure systems. Furthermore, country specific macro-economic and macro-political conditions influence the respective forest sectors.

---

202 BONITA ET AL. (2002: i)

203 PORTER (1990: 742ff.)

204 As compiled by BONITA ET AL. (2002: 28)

Table 6. Major problems of Latin American forest sector clusters[205]

Country	Firm-level	Meso-level[206]
Argentina	Inferior quality of raw material; Lack of co-operation between different sectors; Low technology level; Difficult access to venture capital	Labour legislation; Currency policy; Weak micro-level associations; Lack of participatory forest policies; Poor natural forest management policies
Brazil	No recognition of forest resources as source of employment and income generation in rural areas	Widespread corruption; Historical tradition of deforestation; Unequal land tenure system; Lack of appropriate forest management policies
Chile	Little inter-firm co-operation; No recognition of forest resources as source of employment and income generation in rural areas; No value-adding to forest products; Mainly exports of raw materials	High costs of land (=high opportunity costs of forests[207]); Lack of clear long-term forest management policies;
Colombia	Stagnant forest enterprises; No co-operation between forest enterprises; Lack of co-operation with other industrial sectors; Insufficient technological equipment; Lack of human capital in the forest products sector	Difficult socio-political environment ('Violencia'[208]) ; Lack of external capital inputs; Weak forest management institutions; Lack of public-private initiatives
Mexico	Constraints to raw material supply (resulting in illegal raw material supply); No organised micro-level; Lacking or complicated financing possibilities	Unclear land tenure and land leasing constellations; No clear legal assignment of production and protection forests; Weak and unstable structures in the forest service

205 Based on the conclusions and recommendations of BONITA ET AL. (2002)

206 The term meso-level is introduced and explained in *chapter3.1*. It refers to political and institutional arrangements and actors influencing economic sectors, in this case the forest clusters.

207 See *chapter2.3*

208 See *chapter* 4.2

3. Systemic Competitiveness, Innovations, Actors and Interactions

In the preceding chapters, the theoretical platform upon which this study is based was described. The following chapters focus on the description of the concept of 'systemic competitiveness', which represents the analytical reference system of the conducted research, and its understanding of actors and their interactions. The term 'innovation' is introduced, since innovation is a parameter for successful interaction and development, and is an important element of 'systemic competitiveness'.

3.1 Systemic Competitiveness

The concept of systemic competitiveness provides a framework for the analysis of competitiveness in developing countries, and is meant as an alternative to neo-liberal economic analytical models[209]. According to EßER ET AL., its core statements are[210]:

- The stabilisation of macro-economic conditions and the creation of an economic incentive structure for firms are not the only factors for economic competitiveness.
- The promotion of clusters of firms is the cornerstone of successful industrial development. Consequently, the firms will avail of a collective pool of knowledge and learn collectively.
- The creation of a competitive cluster of firms the collective task of government, NGOs and other private actors. The collective consensus and strategy capacity of these actors is crucial to this process.

The concept of systemic competitiveness covers four analytical levels: the micro, meso, macro and meta-level. The ***meta-level*** relates to "*...development-oriented cultural values, which are shared by a large part of society*". These include social recognition of economic success, general acceptance of ideas of predatory and free-riding behaviour, traditional attitudes towards education, co-operation, trust and mistrust-based patterns to relationships, etc. [211]/[212]. Such structural societal norms affect all members of society.

---

[209] EßER ET AL. (1999: 152)

[210] EßER ET AL. (1999: 152f.)

[211] Meta-level phenomena can in part be described by GRANOVETTER`s theory of 'the strength of weak ties', which describes the role of trust in personal interactions and the consequences for the decision-making. Through his studies, GRANOVETTER introduced the aspect of social interactions in economic action, including that of organisations (GRANOVETTER, 1973 and 1982).

[212] ALTENBURG, HILLEBRAND AND MEYER-STAMER (1998: 13f.)

The meta-level comprises basic elements, which characterise the theoretical concept of social capital. Social capital emphasises the relationship between the members of a society and describes their "*ability to secure benefits through their membership in networks and other social structures*"[213]. These relations are entrenched and governed by a set of elements, such as trust, networks and norms[214]. Social capital comprises elements of the institutional economics and elements of the meta-level, as understood by systemic competitiveness. Therefore, it does not serve with the necessary accuracy the analytical levels provided by systemic competitiveness. Nevertheless, there is a close relationship between these two concepts, and systemic competitiveness is ultimately geared towards investment in and the accumulation of social capital[215].

The ***micro-level*** perspective refers to the firm level. The key actors are entrepreneurs, who face economic challenges in order to achieve competitiveness in the markets. The competitiveness of an economic cluster depends on a firm's ability to "*...offer a variety of products without sacrificing quality and efficiency*", and its responsiveness, which is its ability to "*...react quickly to changes in demand and new opportunities*"[216]. In order to perform according to these premises, the firm must organise its production process, product development strategy and position within the value-chain[217]. The micro-level analysis emphasises two perspectives: the intra-firm perspective and the inter-firm interaction perspective. The latter is considered to be the key to successful economic performance and progress[218].

The ***macro-level*** is the level of the national and international political and economic environment. Unstable or highly biased macro-economic framework conditions[219] reduce the ability of the commodity and capital markets to function properly. This has a negative effect on economic growth and exports. Therefore, the stabilisation of these framework conditions is a crucial element of promotion strategies. The strategies include the reduction of inflation rates and state expenditure[220], the structuring of the fiscal, monetary and budgetary policy, reasonable exchange rate policies and a liberalised[221] trade policy[222].

---

[213] NIELSEN (2000: 19)

[214] NIELSEN (2000: 20ff.);

[215] See NIELSEN (2000:39ff.)

[216] ALTENBURG, HILLEBRAND AND MEYER-STAMER (1998: 9)

[217] EßER ET AL. (1999: 157f.) and ALTENBURG, HILLEBRAND AND MEYER-STAMER (1998: 9f.)

[218] MEYER-STAMER (2001: 2)

[219] ALTENBURG, HILLEBRAND AND MEYER-STAMER (1998: 7) named factors such as high inflation, persistent high budget and trade deficits, anti-export biases, overvalued exchange rates, protectionist barriers, etc.

[220] The reduction of state expenditures should not apply to the basic needs of society; the provision of education, health or infrastructure, but rather to consumption expenditures and expenditures for privileged interest groups (ALTENBURG, HILLEBRAND AND MEYER-STAMER, 1998: 7).

[221] ALTENBURG, HILLEBRAND AND MEYER-STAMER (1998: 9) argue that trade liberalisation may be achieved selectively in the existing economic sectors.

[222] ALTENBURG, HILLEBRAND AND MEYER-STAMER (1998: 7ff.)

Whilst most other approaches and analytical models of economic sector promotion[223] take into consideration the micro and the macro-level, the systemic competitiveness approach also includes the analysis and interpretation of the meta-level. The ***meso-level*** is characterised as the "*level of specific sector policies and institutions between the macro-level, ..., and the micro-level of the firms and the industry*"[224]. The meso-level referred to in systemic competitiveness must be distinguished from the meso-level as it is used in a hierarchical context. It is also distinct from administrative units, such as territorial area levels, from the state down to the smallest administrative unit (for example provinces, departments, federal states, etc.)[225]. At the meso-level of systemic competitiveness, the state, the associations of the micro-level, public-private co-operations and other organised actors create through collective action measures to influence the economic performance of a sector or a business location[226]. Sometimes, a differentiation between macro and meso-level is not possible, as macro-policies and their enforcement usually affect the meso-level, as for example, trade barriers or environmental regulations[227].

As explained in *chapter* 2.2, the concept of economic clusters, defined by MEYER-STAMER and his colleagues, is the theoretical concept of regional agglomerations of economic units, which underlies the concept of systemic competitiveness. The measures initiated by the meso-level actors in order to create and promote the competitiveness of these economic clusters are classified into internal and external incentive structures. The internal structures are: the provision of entities and institutions responsible for the economic performance of the private enterprises, which grant R&D, decentralise economic responsibility and creativity, and enable the entrepreneurs to participate in association. The only external incentive structure referred to by MEYER-STAMER is international development and technical co-operation[228].

According to empirical studies conducted by MEYER-STAMER, EßER and others, the meso-level is the most important of the four levels in the systemic competitiveness concept in industrialised and developing countries[229]. Whereas in developing countries the meso-level commonly seems to be underdeveloped, in industrialised countries the meso-level is frequently overdeveloped, leading to fragmentation and inefficiency[230]. The meso-level actors' interactions lead to sector relevant strategic action. This assumes that the actors are capable of handling collectively, as understood by OLSON, who defined collective action as

---

[223] For example, in PORTER'S competitive advantages in the original industrial districts, in the innovation networks as described by COOKE (1996), and in the various models of regional and national innovation systems (as described by KOSCHATZKY, 2001: 173ff.).

[224] MEYER-STAMER (2001: 2). Translated from German.

[225] MEYER-STAMER (2001: 7) referred to a study by BERGMANN AND FRESER (1999) as an example of the application of meso-level in a territorial-political administrative context.

[226] MEYER-STAMER (2001: 2)

[227] MEYER-STAMER (2001: 7)

[228] MEYER-STAMER (2001: 3)

[229] ALTENBURG, HILLEBRAND AND MEYER-STAMER (1998); MAGGI (2000); MEYER-STAMER (1996b); MEYER-STAMER (2000); MEYER-STAMER (2001b); EßER ET AL. (1995a); EßER ET AL. (1995b)

[230] MEYER-STAMER (2001: 45)

the ability of groups to influence policy and decision-making through self-organisation and the common representation of interests[231].

The concept makes possible the analysis and structuring of economic clusters[232], whereby the analysis and interpretation of the levels provide valuable information concerning the factors for economic growth and development, taking into consideration a wide range of socio-political and socio-economic aspects. The core statement of systemic competitiveness, 'competitiveness through interaction', indicates that the focus of research and analysis must be the actors and their interactions under the given institutional structures. These aspects are, therefore, the research focus of this study. In *figure* 1, the analytical levels and the basic determinants of systemic competitiveness are illustrated.

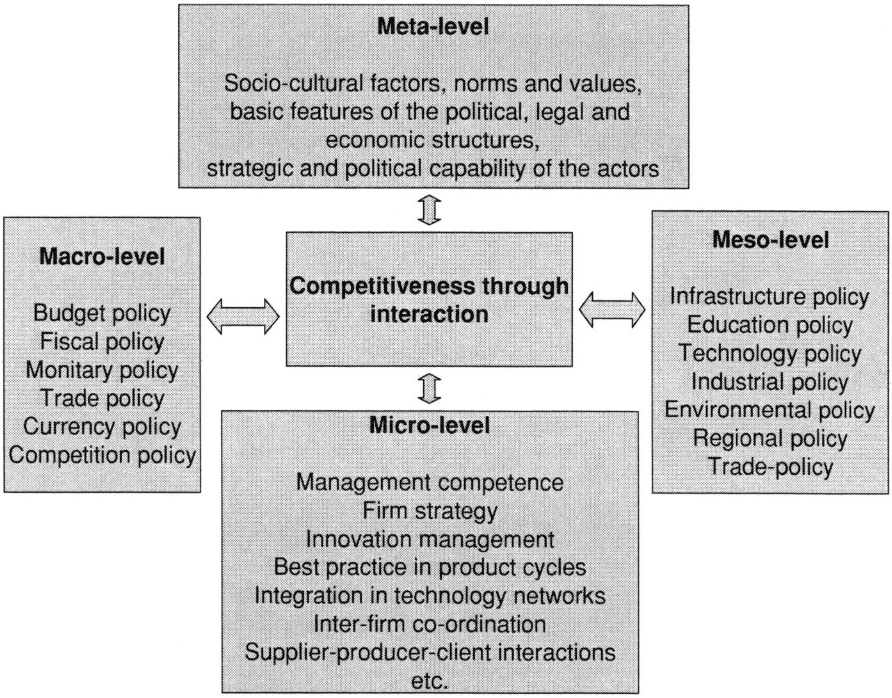

Figure 1. Determinants of systemic competitiveness (Transl. from Eßer et al. 1999)

---

[231] Olson (1998)

[232] Meyer-Stamer (2001: 44f.)

The concept of systemic competitiveness draws from a set of approaches, concepts and theories and combines them in a heuristic framework to analyse factors of growth and development[233]. This theoretical platform is classified into economic and sociological theoretical strands.

The relevant economic theories are constituted by[234]: 1) innovation economics, including evolutionary economics; 2) elements of the post-structuralist discussion; 3) elements of 'New Institutional Economics'; and 4) the management sciences and micro-economics. As was indicated in the preceding chapters, technological progress and its underlying factors are of special interest for the explanation of growth and development. Hence, the theoretical framework of innovation and evolutionary economics is the central research interest of this study. The following paragraphs briefly summarise these three theoretical elements of systemic competitiveness, while innovation and evolutionary economics, and their theoretical fundament, are discussed in detail in *chapter* 3.2.

Post-structuralism[235]: Post-structuralism focuses on the role of the state during the process of industrialisation[236]. This role, in the post-structural sense, is one of strength, with the provision of economic incentive structures to stimulate competitive pressure and technological learning and upgrading.

New Institutional Economics[237]: New Institutional Economics (NIE)[238], which is rooted in the neo-classical growth theories, seeks to identify institutions, which shape the behaviour of economic agents. Institutions, in the NIE context, are rules, conventions, norms and structures. These may be either formal or informal, and are endowed with an enforcement mechanism. Two core elements of the NIE are of particular importance to the concept of systemic competitiveness: property rights and transaction costs. Whereas the importance to economic development of clearly defined property rights is obvious, the term transaction costs requires a more detailed explanation. The transaction costs are costs involved in the arrangement, supervision and enforcement of institutions[239]. Firms consider transaction costs as a decisive factor in decision-making[240].

Micro-economics[241]: The core element of micro-economics and management sciences relevant to systemic competitiveness is the competitiveness of the firms[242]. The sustained

---

233 ALTENBURG, HILLEBRAND AND MEYER-STAMER (1998: 4)

234 As listed by EßER ET AL. (1999: 148ff.) and ALTENBURG, HILLEBRAND AND MEYER-STAMER (1998: 4ff.)

235 As described in ALTENBURG, HILLEBRAND AND MEYER-STAMER (1998: 5)

236 Literature recommended by ALTENBURG, HILLEBRAND AND MEYER-STAMER (1998): AMSDEN (1989); LALL (1992); CEPAL (1990); HILLEBRAND (1991)

237 As described in EßER ET AL. (1999: 149) and ALTENBURG, HILLEBRAND AND MEYER-STAMER (1998: 5)

238 Important New Institutional Economics authors include WILLIAMSON (1995) and NORTH (1988 and 1991)

239 According to NORTH, transaction costs are related to market institutions.

240 See COASE (1937) and NORTH (1988)

241 As described in MEYER-STAMER (2001: 5), EßER ET AL. (1999: 149f.) and ALTENBURG, HILLEBRAND AND MEYER-STAMER (1998: 5)

242 Competitiveness as understood by PORTER. See *chapter2.1*.

competitiveness of a firm is based on four aspects: 1) strategic vision of the firm; 2) demand on the domestic market; 3) capable supporting industries; and 4) the existence of specific supporting institutions. Entrepreneurs depend on these factors in order to maintain profitability and market share. Furthermore, these factors enable entrepreneurs to maintain market-relevant interactions and to remain innovative.

With respect to the part of the theoretical framework of systemic competitiveness rooted in the social sciences, the concept draws upon[243]: 1) economic sociology; 2) industrial sociology; 3) economic geography; and 4) political sciences. For the purposes of this research, an analytical approach provided by the political sciences was drawn upon to investigate actor behaviour and interactions, as established by MAYNTZ and SCHARPF[244]. This approach is described in detail in *chapter 3.3*. *Table 7* lists the most important elements of the sociological-theoretical strands considered in the concept of systemic competitiveness.

Table 7. Socio-theoretical elements of the concept of systemic competitiveness[245]

Theory	Concept-relevant elements
Political sciences	Policy networks and new steering concepts[246]
Economic sociology	Relations of power in economic transactions; Trust and relational contracting; Industrial districts[247]
Industrial sociology	New concepts of production; Systemic rationalisation; Slim production[248]
Economic geography	Agglomerations and other territorial aspects[249]

3.2 Innovation and Evolutionary Economics

The preceding chapters charted the relationship between economic growth and development, examining the role of forest resources in this context. Technological progress has been identified as the major factor behind economic growth. Technological progress underlies the processes of value-adding, which in turn facilitate the improved competitiveness of forest resource-based products in domestic and international markets. "*Technology enables the human being to participate successfully in the natural system and extract its resources*[250]".

[243] As listed by EßER ET AL. (1999: 150f.)

[244] SCHARPF (1993, 2000); MAYNTZ AND SCHARPF (1995); MAYNTZ (1991, 1993, 2002)

[245] Adopted and modified from MEYER-STAMER (2001: 5)

[246] Central authors according to MEYER-STAMER (2001): SCHARPF (1993); MAYNTZ (1991); WILLKE (1988)

[247] Central authors according to MEYER-STAMER (2001): GRANOVETTER (1992); PLATTEAU (1994a and b); SCHMITZ (1995); BECATTINI (1990)

[248] Central authors according to MEYER-STAMER (2001): KERN AND SCHUMANN (1984); WOMACK, JONES AND ROOS (1990)

[249] Central authors according to MEYER-STAMER (2001): STORPER (1997); SCOTT (2000)

[250] WEIZSÄCKER (1990: 14)

Forest resources play an increasingly important role in the considerations of regional and rural development strategies. However, value-adding is crucial to the successful implementation of development strategies integrating forest resource utilisation.

The neo-classical and new growth theorists recognised technological progress in various modifications and versions of the economic production functions of growth. However, the reasons underlying the process of technological progress could not be explained by the economic growth theories. The development theories, on the other hand, have increasingly diverged from the economic growth theories. These theories previously held that development and growth were inhibited by a lack of supply and demand of capital. Today, access to knowledge and capital, as well as the improvement of the human capital assets, are perceived as the major keys to development. Both economic growth and development theories view these human capital assets as crucial to the accumulation of individual knowledge, which is limited by the availability of common knowledge and the cognitive capacity of the individual[251]. Therefore, the goals of growth and development promoting strategies (if aimed at the promotion of human capital) have to be the improvement of the facilitating assets, such as basic and advanced education, health and nutrition status, the facilitation of knowledge exchange and interaction, etc[252].

This chapter features an explanation of how 'technological progress' widens the perspective of growth and development theories through the introduction of the term 'innovation'. It draws on attempts at explanation from evolutionary economics, which emphasise human behaviour and interactions, their complexity and the path dependency of development. The chapter concludes with the presentation of innovation as an indicatorr for successful intra and inter-level interaction.

3.2.1 Innovation and Diffusion

The neo-classical and new growth theories both comprise 'innovation economics', which deals with the phenomenon of technological progress. Innovation depends on driving forces, which the growth and development theories have not been able to explain. Thus, "*the understanding of the innovation process and its economic impact is still deficient*[253]". The concept of systemic competitiveness picks this question as one central theme of its approach. An actor-centred perspective is considered an appropriate means to explain the knowledge gaps. According to systemic competitiveness, innovation is the result at the micro-level of meso-level promotion activities, which have to be interpreted in a meta-level context[254].

---

[251] LACHMANN (2003: 202)

[252] Compare GUNDLACH (1999: 173ff.)

[253] OECD (1996: 5)

[254] MEYER-STAMER (2001: 28ff.)

According to SCHUMPETER, who first examined innovation for economic growth, firms and entrepreneurship, innovations can be found in form of[255]:

- A new product or a qualitative change in an existing product
- A new production process
- Opening of a new market
- Opening or development of new sources for raw materials and other inputs
- The changes to organisational structures

For the purposes of this research project, the term 'innovation' is understood as 1) the innovation of technological products and processes, which according to the OECD are new to the firm and comprise "*technologically new products and significant technological improvements in products and process.*" Thus, innovation comprises also the introduction to the market of an innovative product, or a new production process[256]; 2) an innovation is a significant production and marketing related organisational change, which is new to the firm. In both cases the criteria is 'new', a new product, a new process, a new market, a new supply chain, etc. 'New', in this context, has to be understood as new to the firm (not new to the world), or as a significant improvement[257]. The distinction between new and improvement in innovation research literature is expressed by the use of the terms *radical* innovation for innovations, which are absolutely new to the firm, and *incremental* innovation for innovations, which represent improvements and further developments of already existing products, processes and structures[258].

Research has in recent years come to focus increasingly on the critical importance of the *diffusion* rates of the innovation process; an aspect outside the classical research focus of innovation research, as it has traditionally been conducted by the neo-classical and new economic growth theorists[259]. The process of innovation is only complete, when the innovation has been successfully introduced to a market, or the process or organisational changes have been shown to be effective and efficient. Innovation does not stop with the mere creation or introduction of a new product, process or organisational structure. "*Diffusion is the way in which technological product and process innovations spread through market or non-market channels from their first worldwide implementation to different countries and regions and to different industries/markets and firms*[260]". Whereas the innovation process is

---

[255] According to SCHUMPETER (1934); compiled in OECD (1996: 16).

[256] OECD (1996: 31)

[257] Compare OECD (1996: 31)

[258] Compare OECD (1996: 32)

[259] Based on the R&D dependent models of the "New Growth Theory", as described by ROMER (2001); see also OECD (1996: 5)

[260] OECD (1996: 9)

of importance for the sustained competitiveness of individual firms, the diffusion rate of innovations is crucial for the competitiveness of an economic sector as a whole.

The diffusion of innovation has been investigated since initial studies by SCHUMPETER in the 1930s. In 1962, ROGERS concluded that the adoption of a technological innovation by entrepreneurs depends on five factors: 1) the relative advantage, which can be measured in economic terms, but also in terms of social prestige and satisfaction[261]; 2) the compatibility with others, who are interlinked or competing with the adopter; 3) the complexity, which limits adoptability and applicability; 4) the possibility to conduct trials and modify the adoption; and 5) the ability to monitor the innovation over time. ROGERS developed an adopter classification, which comprises five types: the innovator; the early adopter; the early majority adopters; the late majority adopters; and the laggards (see *figure* 2).

This classification recognises the fact that innovations are not simultaneously adopted by all, or even the majority of actors. Instead, the adoption of new products, processes, etc. takes place in a process, which is influenced by the restricted rational of actors, by structural factors and by socio-psychological factors[262]. Hence, "*diffusion is a process by which an innovation is communicated through certain channels over the members of a social system*"[263]. This means that actor interactions are crucial to the innovation diffusion process within economic sectors. Thus, the diffusion process is closely linked to basic features of the meta-level, which provide the platform for inter-actor interaction and knowledge exchange. It is also dependant upon the human capital features of the individual. Generally speaking, social capital enhances innovation and innovation generates economic growth and development[264].

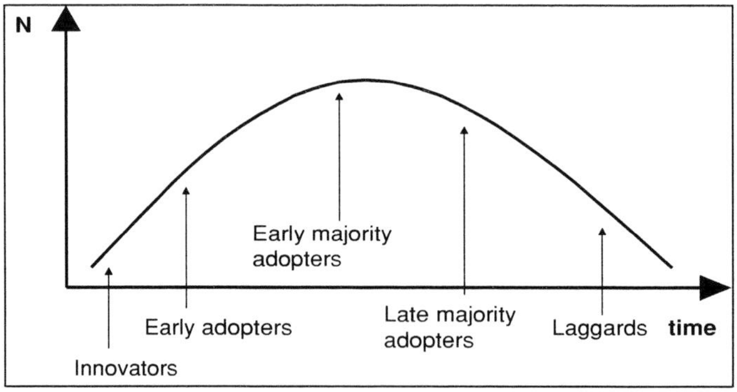

Figure 2. Adoption cycle amongst members of a social system (According to Rogers, 1995)

261 BOUDOURIDES (2001: 32)

262 As concluded by RYAN AND GROSS in 1943.

263 ROGERS (1995: 5)

264 NIELSEN (2003: 2)

Explanations of the diffusion processes have been attempted using models, principally '*threshold models*' and '*critical mass models*'[265]. Basically, the threshold models declare that an adopter estimates the personal benefits and costs of adoption and only does so when the perceived benefits exceed the perceived costs[266]. A threshold can also be related to the proportion of adopters within the personal network of potential adopters. Their number is decisive for the individual's decision. The critical mass models refer to the minimum number of users within a system, which is crucial to sustain the diffusion process of an innovation. It assumes that each adopter is the reference point for a number of further adopters. Important in this context are adopters of high centrality, meaning of high importance, such as large-scale producers or firms, or actors of good repute[267]. There are a number of diffusion channels for technological innovations, of which the following are only a sample[268]:

Diffusion takes place through ...

- fluctuation and movements in the labour market, i.e. when skilled workers change occupation or employer.
- new firm entries and newcomers, i.e. new firm entries of former employees.
- sub-contracting and demand, i.e. suppliers of advanced national and international firms.
- international technical co-operation and development programs.
- R&D; i.e. in special R&D departments or sections of firms.
- imitation, i.e. product imitation.
- spill-overs, which are externalities of tacit individual knowledge, which become commonly available through market exchange, co-operations and the copying of products.
- the dissemination of common knowledge, i.e. in schools and similar education systems.

This study does not involve an investigation into entrepreneurship, and only partly into entrepreneurial behaviour. An exhaustive integration of these aspects would open a wide set of new perspectives and theoretical approaches. However, the entrepreneur and his decision-making behaviour are inseparably linked to the innovation and diffusion process. In the

[265] BOUDOURIDES (2001: 33)

[266] GRANOVETTER (1978: 1422). For more literature on threshold models in diffusion processes see the publications by GRANOVETTER AND SOONG (1983, 1986 and 1988).

[267] BOUDOURIDES (2001: 33); Authors focusing on the critical mass approach were: OLIVER AND MARWELL (1988) and PRAHL, MARWELL AND OLIVER (1991)

[268] Combined listing according to ELIASSON (2000; 2001:154) and BRENNER AND GREIF (2003)

context of this study, the entrepreneur is seen as an actor at the micro-level of the systemic competitiveness concept, who behaves according to the patterns of micro-economic decision-making and social interaction. The micro-economic understanding of the entrepreneur as an autonomous leader of one or several firms, handling in accordance with the basic principals of (neo-classical) economics, constitutes one aspect of the interpretation context of the research. The classical entrepreneur leading a firm acts according to the following patterns[269]:

- The firm is an organisational unit
- The production unit (firm) uses the combination of the production factors capital, labour, information and production inputs
- It produces to meet the demand of market participants
- The firm constitutes part of the market and social environment
- Production is organised according to economic principals[270]
- The entrepreneur is profit oriented
- The entrepreneur seeks to balance the finances in order to maintain liquidity

These characteristics directly determine the criteria according to which the entrepreneur adopts innovations. The assumption is that innovations are the direct result of the conscious R&D activities of the entrepreneur, which enable him to improve the value-adding processes and to improve competitiveness[271]. On the other hand, innovation increases competition between firms, which in turn results in stronger R&D activities and more elaborated innovations to meet this increased competition[272].

Besides the diffusion processes taking place within an economic sector or a society, the inter-regional patterns of the diffusion of innovations constitute an important aspect of economic growth theory and its innovation research. The regional comparison of different degrees of innovation are usually described through 1) the number of patents registered in a region and its economic sectors; and 2) through the intensity of R&D activities the members of an economy and the whole national economy are undertaking, which is measured either through the investments firms and entrepreneurs effect in their R&D sections, the number of R&D staff members or the percentage of the national GDP invested in R&D activities[273].

---

[269] According to SCHMITHÜSEN ET AL. (2003: 97)

[270] The basic economic principals are known as the principals of 'maximum output' and 'minimum input'. See LACHMANN (2003: 10)

[271] SCHMITHÜSEN ET AL. (2003: 125)

[272] SCHMITHÜSEN ET AL. (2003: 125)

[273] BATHELT AND GLÜCKLER (2002: 227ff.). The ratio R&D/GDP is known as the GERD ratio in economic data analysis.

FERRANTI ET AL. compared Latin American and East Asian countries with respect to the number of patents as an indicator of innovation activities and found significant differences[274]. The number of patents per million inhabitants in Latin American countries is only 2.07, compared to 18 patents in East Asian countries. International patents are even less common in Latin American countries. For example in Latin America only 0.8 U.S. patents are registered per million inhabitants, in East Asia the number is 54.4.

LEDERMAN AND MALONEY (2003) revealed a correlation between R&D and national GDP. They argued that in developing countries investments in R&D activities account for higher returns, although the starting costs are comparable to those of industrialised countries. Furthermore, they found that in developing countries with abundant natural resources the R&D returns are comparably higher with lower overall investments[275]. However, in developing countries significantly less of the GDP is spent on R&D related activities. The sources of investment in R&D are approximately 40-50% from private enterprises, 40-50% public expenditure and around 10% from other sources, amongst them the Foreign Direct Investments (FDIs)[276]. These figures may vary significantly from one country to the other. In developing countries an average of 1,8% of the annual GDP has been spent on R&D in the last 10 years, while in the industrialised countries an average of 2,5% of the annual GDP has been spent on this purpose in the same period[277]. Investments in R&D in the developing countries have mainly been made by private enterprises, while the role of the state has decreased. FDIs play an important role in financing R&D in developing countries. However, they vary from almost 0% in African countries to more than 60% in Panama[278].

Both figures, the registered patents and the extent of R&D activity, may be used to describe the state of innovation within a nation or a sector from a neo-classical point of view. They may also be entered as variables in the factor productivity, and explain the performance of an economy. However, these factors ignore the innovative activities of large numbers of actors and entrepreneurs in developing countries, as they are informal micro-enterprises without resources or access to patent registration and R&D investment. Nor are the individual skills, as well as the knowledge diffusion processes and the factors underlying them, considered[279].

Geographical economics has attempted to explain the diffusion processes between regions, mainly through the following approaches: spatial diffusion, temporal diffusion, and space-time-related explanation approaches[280]. Spatial diffusion depends upon the physical proximity

---

[274] FERRANTI ET AL. (2003: 42)

[275] LEDERMAN AND MALONEY (2003)

[276] As compiled from UNESCO database analysis (www.uis.unesco.org/TEMPLATE/html/Exceltables/science/GERDMay2004.xls) in 2004.

[277] Calculated from the last available figures in: www.uis.unesco.org/TEMPLATE/html/Exceltables/science/GERDMay2004.xls in 2004.

[278] As compiled from UNESCO database analysis (www.uis.unesco.org/TEMPLATE/html/Exceltables/science/GERDMay2004.xls) in 2004.

[279] See also BATHELT AND GLÜCKLER (2002: 234) and FERRANTI ET AL. (2002: 49ff.)

[280] Compare BATHELT AND GLÜCKLER (2002: 231f.)

of regions. This proximity facilitates or blocks the personal or inter-firm contact, and with it the possibility of spill-over effects and active knowledge and technological exchange. The temporal diffusion process relates to diffusion within an economic sector (as shown in *figure 2*). The mixed space-time diffusion approach emphasises the importance of both aspects. The numbers adopting an innovation decrease with increasing physical distance, resulting in a wave shaped diffusion pattern, as originally described by MORRILL (1968). This mixed diffusion model also explains the interrelations between different innovations within an economy, which compete for resources within and adopters[281].

3.2.2 Innovations and Evolutionary Economics

While neo-classical approaches explaining technological progress and innovations are based on the collocation of these factors within the equilibrium of the production function[282], evolutionary economics has increasingly emphasised the importance of societal complexity and the role of actor behaviour and interaction in the innovation process[283]. The neo-classical innovation approach has been subject to growing criticism, which culminated in declarations stating that[284]: 1) the cognitive and learning processes of economic subjects (firms, organisations, entrepreneurs, etc.), and their expectations and preferences, are not considered sufficiently in the neo-classical approach; and 2) the firm dynamics caused by processes in the market are not considered either.

In *table 8* the basic patterns of the neo-classical and evolutionary economics approaches in explaining technological progress and innovation are compared. Central to the evolutionary economics approach is the entrepreneur as the actor, and the innovation process itself. Evolutionary economics has its origins in SCHUMPETER'S benchmark work of 1934, '*Theorie der wirtschaftlichen Entwicklung*'. SCHUMPETER described the entrepreneur as the driving force behind technological progress, which in turn is the starting point of economic development. He considered technological progress to be an endogenous factor of economic growth, which described the qualitative improvement of an economy[285]. According to this understanding, the engine of innovation is the entrepreneur. Hence, it is not induced by the demand structures of the market as it is in the neo-classical economics. Innovation according to the evolutionary, SCHUMPETERIAN understanding is achieved through the '*technology push*', while the neo-classical theory makes the market and its '*demand-pull*' responsible for the creation and adoption of innovations[286]. This contradicts the neo-classical assumption, which is inherent in the individual rational choice model of the '*Homo oeconomicus*[287]', that

[281] BATHELT AND GLÜCKLER (2002: 233)

[282] As established by SOLOW.

[283] BOLLMANN (1990: 18ff. and 40ff.)

[284] KOGLER (1991: 72)

[285] BOLLMANN (1990: 43ff.)

[286] Compare FREEMAN (1974: 164)

[287] See *chapter 3.3*

any change of behaviour is a consequence of a change of conditions, premises and incentives; all innovation originates outside the system[288].

NELSON AND WINTER (1975, 1977 and 1982) adopted the evolutionary perspective of economic growth processes and technological progress[289]. They laid the foundations for the increasing influence the evolutionary perspective has gained in innovation research over the last decades through the application of computer-based models aimed at the description of the complex patterns of evolutionary innovation economics[290]. The theoretical strand derived from their idea of combining apparently unrelated factors in the explanation models of economic growth and innovation has become known as '*New Evolutionary Economics*'[291].

'*New Evolutionary Economics*' draws upon behavioural and cognitive theories and introduces the factors insecurity and complexity into the economic decision-making process[292]. The role of system complexity is considered one decisive factor in innovation and diffusion processes from the evolutionary perspective. The basic assumption of complexity is that technological progress can not be rationally or administratively induced and steered, as too many coincidences and too much stochastic instability affect the individual innovation and adoption processes[293].

Complexity basically refers to: "...such things as the size of society, the number of and distinctiveness of its parts, the variety of specialised social roles that it incorporates, the number of distinct social personalities present, and the variety of mechanisms for organising these into a coherent, functioning whole"[294]. According to this assumption, the only way to promote innovation and diffusion is to increase the number of adopting actors, which means to increase the critical mass and to integrate actors of high centrality[295]. The complexity of society, economy, the firm, etc. results in an increased diversity of individual knowledge and in a number of possibilities for the individual to increase existing and gain access to new knowledge. Hence, the evolutionary theorists have introduced the term the 'experimentally organised economy'[296]. This is the evolutionary-complex description of what traditional economists termed the 'knowledge economy'[297].

---

[288] LOASBY (2000: 298)

[289] For a detailed description of '*New Evolutionary Economics*' and their criticism of neoclassical growth theories see NELSON AND WINTER (1982: 207ff.)

[290] ANDERSEN (1996b: 101ff.)

[291] ANDERSEN (1996a: 10f.)

[292] KOGLER (1991: 7)

[293] EBELING AND SCHARNHORST (2001: 59f)

[294] TAINTER (1988: 23). Complexity is encountered through the self-organisation processes within an economy, as organisation promotes competitive and other advantages (See EBELING AND SCHARNHORST, 2002, BRENNER, 2000).

[295] EBELING AND SCHARNHORST (2001: 59ff.). This was shown for German clusters by BRENNER AND GREIF (2003).

[296] See CARLSSON AND ELIASSON (2001)

[297] See ELIASSON (2001: 42ff.)

This is the point where evolutionary economic innovation theories contradict the traditional explanation models. The neo-classical measures for promoting economic sectors are not flexible enough to meet the complexity of the micro-level, as they are frequently based upon macro-level growth models[298]. Furthermore, these approaches promote firms and industries through subsidies and incentives, resulting in 1) the loss of skilled labour to low productivity and low technology industries, labour that could be used in more progressive and promising industry branches; and 2) the artificial maintenance of non-profitable firms and industrial branches: the '*losers*'[299]. Two basic aspects have to be considered as main keys in the explanation of technological progress from an evolutionary point of view: 1) the path dependency or the trajectories of innovation/technological progress and 2) the role of tacit and common knowledge and their development[300].

Table 8. Major differences between neo-classical and evolutionary understandings of innovation[301]

	Neo-classical Economics	Evolutionary Economics
Basic inputs for innovation	Production factors (hard and soft); R&D; Competition	Pool of common knowledge; Tacit knowledge
Origins of innovation	R&D	Cognitive capacity of actors; Actor behaviour; Learning organisations
Indicators for measurement of innovation	Expenditures for R&D (input); Employees in R&D (input); Patents (output); Growth rate (indirectly via the production function)	Output productivity
Driving forces of diffusion	Competition	Interaction
Development model	Product cycle	Path dependency; Trajectories

[298] ELIASSON (2001: 58)

[299] CARLSSON AND ELIASSON (2001: 17f.). The winner and loser concept in economic development was described by NELSON AND WINTER (1984).

[300] For the discussion of these both aspects see *chapter 3.2.3*.

[301] Compiled according to BOLLMANN (1990) and KOGLER (1991)

3.2.3 Innovation Systems

The theories of economic growth have adopted the regional context as the system most crucial for a firm's or an economic sector's performance[302]. A region is "*...a rather stable homogenous entity in terms of its collective knowledge, institutional structures and social conventions*[303]". KOSCHATZKY viewed the region as a research-oriented unit, which can be defined according to aspects of geographical homogeneity or economic functionality or administrative-political aspects, etc[304].

In the literature, the *national innovation systems*, the *regional innovation systems* and the *regional innovation networks*[305] are the most common systemic views on economic innovation processes. In the 1980s and 1990s the concepts of national and regional innovation systems were developed based on the evolutionary perspective of innovation processes[306]. The focus has increasingly been put on the regional aspects, which emphasise the role of proximity between innovating actors resulting in specific regional patterns of innovation[307].

In a '*National Innovation System*' (NIS), public policies, the R&D activities of private enterprises, the industrial culture, and education are to be co-ordinated and linked[308]. Of special importance for the progress of innovation and diffusion in the NIS are the interactions between the public and the private actors[309]. According to KOSCHATZKY the most important features of the NIS are[310]:

- The institutional structure of a country, which includes the public and private organisations, and the existence of institutions to promote and regulate technological change.
- The incentive structures of a country, which are clearly defined institutions to enhance technological change, education, entrepreneurial activities and mobility of labour.
- The creativity of the economic actors in a country, which depends strongly on the human capital resources.

---

302 See KRUGMAN (1991: 1) and KOSCHATZKY (2001: 1). See also *chapter 2.1* of this study on the agglomeration of firms.

303 BOSCHMA AND LAMBOOY (1999: 415)

304 KOSCHATZKY (2001: 15)

305 The term 'network' is used inconsistently in this study. It has not been used to mean the sociological network concept, as for example by WEYER (2000: 1ff.). Nevertheless, network in this study refers to the links between actors, which must be based on mutual advantage, which is the reduction of uncertainty and the possibility to increase or improve individual performance (WEYER, 2000: 10). A detailed discussion of the role of networks in innovation research is available in KOWOL AND KROHN (2000) and KOSCHATZKY (2001: 137).

306 See NELSON (1993) and LUNDVALL (1992)

307 KOSCHATZKY (2001: 5f.)

308 As compiled by KOSCHATZKY (2001: 160) according to FREEMAN (1987).

309 KOSCHATZKY (2001: 160)

310 According to KOSCHATZKY (2001: 162)

- The cultural characteristics of a country, which are basically the features of the meta-level[311].

While the emphasis of the early NIS research was on the institutional structures, more recent studies have focused on learning and the accumulation of knowledge[312]. Learning in this context refers to activities leading to technical and economic information, which requires interpretation to be used in decision-making[313]. Knowledge accumulation is based upon the evolutionary perspective of knowledge, made up of two elements, the individual '*tacit*' knowledge, which is comparable to the human capital, and the common '*collectively codified*' knowledge available to all actors in a nation, region or sector[314].

The knowledge required for innovation is a mixture of both, tacit and codified knowledge. McKelvey wrote that while codified knowledge is limited by the availability of information and educational facilities, and the overall state of technology, tacit knowledge very much depends on the individual's abilities, experiences and cognitive capacity[315]. Hence, tacit knowledge can only be shared to a certain degree, and learning is subject to the interactions between the actors. Through the interaction process the actors' tacit knowledge increasingly becomes codified collective knowledge.

Codified collective knowledge, on the other hand, may be adopted by the individual and transformed into tacit knowledge through individual modification and amplification. The standard of codified collective knowledge determines the so-called '*trajectories*' of innovation[316]. These trajectories are an attempt to explain why 1) technological developments in a sector or region tend to concentrate around similar possibilities of alternatives[317]; and 2) why induced innovation processes from outside tend to fail or only partly succeed ('*resistance to innovation*')[318]. The innovation trajectory approach of the evolutionary innovation economics underlines that innovation processes are '*path-dependent*'. This means that technological decisions made in the past influence and determine the decisions and adoption behaviour of the actors in the present[319].

The concept of '*Regional Innovation Systems*' (RIS) was influenced by a set of theoretical developments in the 1990s, such as Porter's competitive advantage, network economics,

---

[311] See *chapter 3.1*

[312] Koschatzky (2001: 163)

[313] McKelvey (1998: 161)

[314] McKelvey (1998: 162)

[315] McKelvey (1998)

[316] McKelvey (1998: 162)

[317] The phenomenon was investigated by Dosi, Pavitt and Soete (1990), when they were comparing international and inter-sector differences and determinants of the innovation process (1990: 73ff.)

[318] See Meyer-Stamer (2001:31ff.)

[319] Compare Meyer-Stamer (2001: 32) and Bathelt and Glückler (2002: 239f.)

agglomeration studies and 'New Institutional Economics'[320]. A '*system*' in this context consists of elements and their mutual interactions, which enable them to produce, diffuse and develop new economic knowledge[321]. An '*innovation system*' is[322]:

- A social system, in which innovations are the result of social interactions between economic actors
- An open system, which is in exchange with its environment

Important elements of a RIS are: the private economic actors, universities, other research entities, public and private finance institutions, education and dissemination entities, policy makers, etc. One important conclusion derived from the study of national and regional innovation systems is that the actors of an economic sector are embedded in a regional context; "*...firms and other organisations act context-dependent and path-dependent*"[323]. Within the RIS concept a number of explanations for the driving forces behind the innovation process have developed, all of them are linked to the interactions between the actors. The most common are the industrial district and the economic cluster approaches[324]. In these approaches, innovation is seen as the result of the interaction of the elements of a socio-technical system (as shown in *figure 3*). MANSKE AND MOON'S socio-technical system is based on the assumptions that innovation results from the interaction of individual and collective (complex) actors [325]:

- "These actors are bound up in structures that enable them to engage in activities, while at the same time imposing constraints on the options available to them."
- The assumed innovation networks are constituted of different, relatively autonomous actors and both formal and informal institutions.
- The innovation process is recursive, which means that the innovation process does not occur in a straight line, but features many backward loops, iterations and overlaps.
- Innovation involves collective learning, which is the combination of collective and tacit knowledge under social, technical and economic considerations.
- Innovation requires the co-operation of meso-level actors to co-ordinate and optimise, and to exchange tacit scientific knowledge, including interaction with the micro-level.

---

320 KOSCHATZKY (2001: 175)

321 COOKE (1998: 11)

322 According to KOSCHATZKY (2001: 175)

323 BOSCHMA AND LAMBOOY (1999: 415)

324 See *chapter 2.2*. For more innovation dynamics approaches see KOSCHATZKY (2001: 184ff.)

325 According to MANSKE (in prep.).

For an analysis and evaluation of these processes MANSKE recommended the application of actor-centred approaches, such as SCHARPF`S and MAYNTZ`S actor-centred institutionalism[326]. The innovations are seen as indicators of the interaction processes at the various levels, as intermediate variables to assess the socio-technical processes in such a system[327].

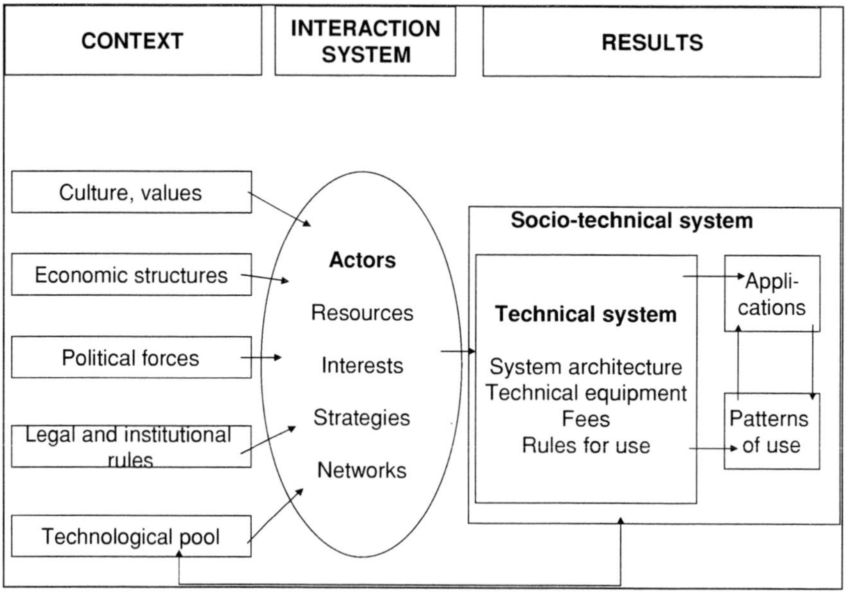

Figure 3. Determinants of innovation in a socio-technical system (According to SCHNEIDER AND MAYNTZ, 1995 and MANSKE AND MOON, 1998)

3.2.4 Innovation in Developing Countries

According to research by ALTENBURG AND MEYER-STAMER in various economic clusters in developing countries, there is a shortage of innovative enterprises in these countries. This can be attributed to a weak or not existent micro-meso interaction policy and low inter-firm interaction. This in turn results in a low degree of specialisation[328].

Innovation processes in developing countries have mainly been investigated in the context of agricultural and industrial sector development[329]. The agricultural sectors were investigated in relation to their capacity to adopt technology introduced from industrialised countries. This

326 See *chapter 3.3*

327 Compare NIELSEN (2003: 2)

328 Compare ALTENBURG AND MEYER-STAMER (1999: 1709f.)

329 See for example LÜCKE (1992) and compare *chapter 2.2* on cluster research.

strategy of induced innovation was very common during the development decades of the 1970s and 1980s. This was an example of development as a process to be enhanced and enabled through external capital and technology inputs.

The researchers, who evaluated these innovation processes in developing countries, concluded that farmers adopt new technology according to their personal financial and intellectual capacity, and their experience. This means that adoption takes place along a gradient between the two extremes '1 to 1 – adoption' and 'non-adoption'[330]. Another conclusion was that the decisions to adopt various innovations are interrelated, as many innovations are introduced simultaneously. The adoption of the various innovations takes place step-by-step. Therefore, one innovation step is frequently built upon by the next[331].

Research into innovation in the industrial sectors in developing countries in the 1980s revealed three main problems facing induced innovation strategies, as were mentioned already[332]: 1) some developing countries have imposed highly restrictive import regimes for machinery, in order to protect domestic suppliers; 2) investment in developing countries is often allocated to enterprises on a discriminatory basis, e.g. in favour of traditional, labour intensive industries; 3) state policies in developing countries frequently tend to raise the cost of fixed capital and may, therefore, impede the adoption of innovations in particular countries or industries; 4) technology and innovations transferred from industrialised to developing countries often exceed the capacity and availability of human capital in the target countries; 5) the efficient adoption of new technology and the education of human capital takes time. The technology and income gap between the industrialised and developing countries increases, as the productivity gains in the industrialised countries are higher as a result of the efficient application of these technologies, while in developing countries a period of learning and education is required.

Probably the most important finding of the early innovation research in developing countries was that the adoption rate of new technology depends strongly on the regional, cultural and social situations of the farmers and entrepreneurs. These are factors beyond the purely economic, and had not been sufficiently investigated until then. Furthermore, in the agricultural context, the adoption of innovations depends on the socio-economic status of the farmers. This results in adoption patterns, which according to ROGERS' adoption cycle (*figure* 2), see the wealthier farmers belonging to the early adopters, whereas the poorer farmers are late adopters or laggards. The wealthier farmers benefit more from the introduction of new technology than the poorer ones, as they are able to generate income more rapidly as a result of the advantages of the new technology[333].

---

330 FEDER (1981: 47)

331 FEDER (1981: 48)

332 As compiled by LÜCKE (1992: 10)

333 Compare FEDER (1981: 48f.)

A study carried out in 1980 by COOPER into policy interventions for technological innovations summarised the particular problems of developing countries with respect to the innovation process as follows[334]:

- The firms in developing countries have a relatively short planning horizon
- The firms put too much emphasis on relatively minor product changes and do not invest in process innovations
- R&D activities are mainly conducted by larger firms

COOPER recommended strong external involvement and the provision of the deficient factors, such as the substitution R&D with imported products and processes, and the contracting of foreign staff and enterprises for setting up firms and designing technologies. On the other hand, he also recognised the problems such a policy would have on the local (firm) population[335]:

- The bypassing of local labour and their exclusion from learning processes in new technologies and enterprises
- The artificial bridging of local innovation 'paths', which results in the loss of knowledge and potential
- The under-use of local innovative skills results in insufficient establishment of local R&D facilities
- The general marginalisation of science and technology in developing countries
- Development backlogs are not tackled as the 'solutions' are provided from outside
- Aspects hindering the development (innovation) of the socio-economic or cultural structures are artificially maintained

More recent research into innovation processes in developing countries is rather scarce. The existing literature focuses mainly on the R&D activities of different industrial and agricultural sectors[336]. Some authors investigated the role of foreign technology inputs on firm productivity and performance[337]. There are only a few publications, which adopted a more systemic point of view, such as the investigation of innovation clusters in Latin America by BORTAGARAY AND TIFFIN in 2000, or the various cluster researches undertaken by MEYER-STAMER and colleagues. There has in recent years been an attempt to transfer the experiences gleaned from innovation research in industrialised countries onto developing countries. The

[334] Compiled according to COOPER (1980: 5f.)

[335] COOPER (1980: 6ff.). In his study, COOPER focuses on the advantages and disadvantages of technology transfer, whereas this study seeks to investigate local development paths. Nevertheless, COOPER laid the foundation for a critical view of external innovation inputs and the increased consideration of local potential.

[336] See for example BEINTEMA, ROMANA AND PARDEY (2000), LEDERMAN AND MALONEY (2003).

[337] See for example KRAAY, SOLOAGA AND TYBOUT (2002), SCHIFF AND WANG (2002).

major analytical levels are the national economies under the considerations of the NIS concept[338]. In general, innovation related research has adopted the basic assumptions of the 'New Growth Theory' that a liberalised market economy enhances the transfer and diffusion of technology and knowledge, that market competition is the driving force behind innovation, and that entrepreneurship and capital endowment are the determining factors of innovation processes. Therefore, the human capital has to be strengthened, in order to facilitate knowledge accumulation and access to capital assets. The general opinion is that state intervention, foreign direct investment and technology transfer are the major driving forces behind innovation, which gives rise to a debate over how much intervention is necessary to enhance economic growth[339].

Apart from this perspective on innovation, there have also been studies of the culture of entrepreneurship and co-operation in several Latin American countries[340]. This examination of the economic effects of meta-level phenomena provided two major insights: 1) the important role of mistrust in social, as well as economic co-operation and interaction[341]; and 2) the important roles played by the central actors in society, as reference actors and role models, as well as steering actors[342]. URREA ET AL. summarised the general problems in Latin America and the obstacles to advanced innovation processes. They stated that[343]:

- There are gaps and breaks in the innovation paths due to colonisation, and the loss of indigenous knowledge.
- The protectionist periods of most Latin American economies have hindered the international diffusion processes and inhibited the expansion of the technology sectors.
- The focus of Latin American innovation promotion is on R&D and promotion of the export sector.
- There is a general research dissemination gap. Scientific knowledge is not disseminated and, therefore, is not applied. Knowledge is lost.
- In general, the technological pool in Latin American countries is very broad, but not advanced.
- The entrepreneurial culture used to be an elitist culture, including the capacity to innovate[344].

---

[338] See for example POTI AND BASILE (2000), DUMONT AND TSAKANIKAS (2001), FERRANTI ET AL. (2003).

[339] For example in the Colombian case, KALMANOVITZ (2001a: 175ff.). For the debate on the role of the state in the economy and development in Latin American countries see FUHR (1993) and SCHNEIDER (1998).

[340] HOUSE ET AL. (2002); OGLIASTRIE (1998); URREA ET AL. (2000)

[341] Compare the publications on trust and interactions by GRANOVETTER and LEVIN AND CROSS (2002).

[342] See OGLIASTRIE (1998)

[343] Compiled according to URREA ET AL. (2000: 70ff.) and their three regional case studies of the Colombian regions of the Valle de Cauca, the Andine region and the Caribbean.

[344] I.e. in URREA ET AL. (2000: 49f.)

- Today's entrepreneurs are mainly informal SMEs, with a low capacity for innovation. There is high fluctuation of firms and only a short-term entrepreneurial horizon.
- Innovation is increasingly dominated by the agricultural sectors and their research entities[345].
- Although in the Latin American context the regional aspects dominate entrepreneurial behaviour, the socio-cultural situation and innovative capacity, the concept of NIS is preferred over RIS approaches[346].

3.3 Actor-centred Institutionalism

When describing the systemic competitiveness approach and the theories on innovation, the importance of the actors, their interaction and the existing institutional structures was highlighted. The meso-level actors were challenged with enhancing and promoting an environment, which supports the micro-level entrepreneurs in gaining competitiveness in national and international markets. *Figure* 4 illustrates typical actors at the different analytical levels in the concept of systemic competitiveness. These actors are linked through communication and networking.

In this concept, actor action is determined by a set of meta-level structures, such as societal values and belief systems, and the institutional arrangements in place. Basically, the concept is derived from MAX WEBER, for whom the actions of an individual are 'purpose-rational'[347]. This definition of actors as individuals acting intentionally within their rational builds on two classic actor models: the *Homo sociologicus* and the *Homo oeconomicus*. Whilst the latter was introduced by the economic sciences and has been drawn upon by the classical and neo-classical growth theories, the *Homo sociologicus* was introduced and applied by sociology and related subjects[348].

---

[345] I.e. in URREA ET AL. (2000: 251ff.)

[346] I.e. in URREA ET AL. (2000: 226ff.)

[347] See WEBER (1922: 12)

[348] SCHIMANK (2002: 20f.)

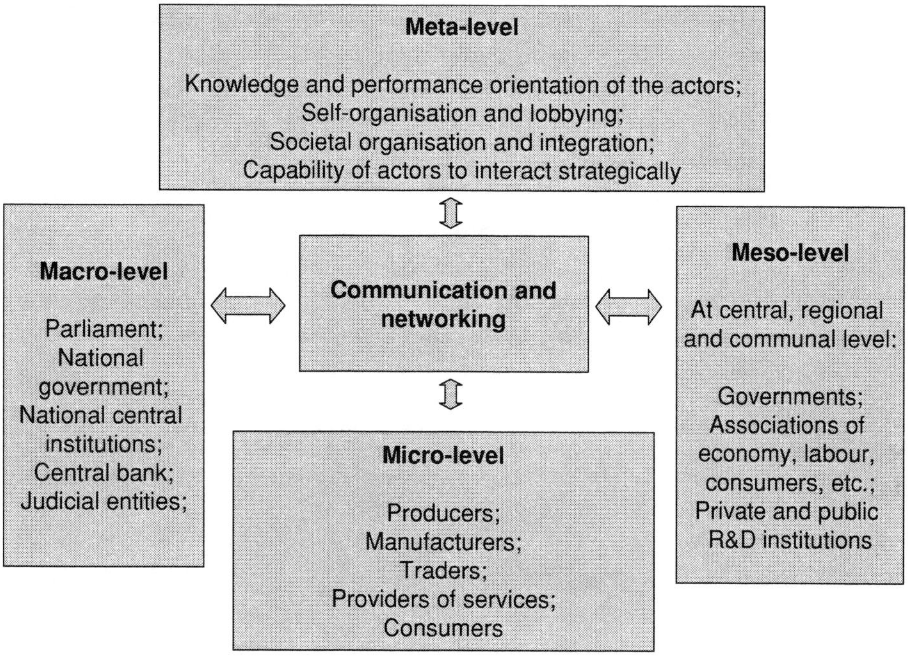

Figure 4. Strategic actor capability (Transl. from EßER ET AL. 1999)

According to the *Homo oeconomicus* model, action is benefit-oriented. Action is clearly aimed at the maximisation of profit and follows the principle of 'rational choice'[349]. 'Rational choice' is characterised by the individual's pursuit of profit given scarce resources, decreasing marginal profitability and opportunity costs resulting in subjective cost-benefit calculations and discounting for the future[350]. Since this actor model is rooted in the neo-classical economic growth model, perfect market and the capital orientation of the production function are assumed[351].

In the case of *Homo sociologicus*, actions are norm-oriented, which leads to role-model-like patterns of behaviour shaped by the social norms and values[352]. The actions of this actor type are characterised by subjective perceptions and expectations. His behaviour is not driven by the pursuit of profit maximisation, but by self-determined and over-directed social factors. These factors are the position of the actor in the social hierarchy and his/her role in it[353].

---

[349] SCHIMANK (2002: 72f.)

[350] SCHIMANK (2002: 73ff.)

[351] See *chapter* 2

[352] SCHIMANK (2002: 38f.)

[353] The role of the actor is also self-determined and over-directed, described as role-making and role-taking processes (SCHIMANK, 2002: 55ff.)

Generally, the motivation behind *Homo sociologicus* is to act towards the resolution of any conflicts emanating from disagreements between himself and other individuals, as well as internal conflicts. His major pursuit is to secure the expected outputs of his perceptions and expectations[354]. The *Homo sociologicus*, in pursuing the resolution of any conflicts, may also act in a profit maximising way, as referred to in the *Homo oeconomicus* model, but the motivation is different.

Neither actor model can on its own fully explain the patterns of behaviour and the interaction of actors in society[355]. Therefore, WEBER combined their basic features, the optimisation of profit and norm-oriented behaviour. The *Homo sociologicus* is considered the prevalent model in attempting to explain the forces driving the behaviour of individuals. *Homo oeconomicus*, on the other hand, is viewed as a model useful for the understanding of the societal processes of development and growth[356].

The concept of systemic competitiveness is based on the premise that actor interaction is the driving force behind competition. Actor interaction requires alternating, as well as weighted and considered action. WEBER distinguishes social action from mere behaviour and non-social action[357]. The basic criterion of social action is conscious and intentional action, which is directed towards another actor. The mutual action then, results in a social relationship, which can be manifested through alternating expectations[358]. The social acting of individuals is the precondition for actor interaction in terms of systemic competitiveness and innovation processes, considering that the reality and rationality of the actors is limited[359].

Based on this understanding of actors, their social behaviour and their conduct, SCHARPF (2000) developed a classification of actors and an explanation model of strategic action and interaction, which is also drawn upon by the concept of systemic competitiveness. The 'actor-centred institutionalism[360]' approach focuses on the power of social interactions to explain societal and economic processes, which take place within institutional structures[361], which in turn affect the outcomes of these interactions[362]. This analytical approach is recommended by the proponents of systemic competitiveness and cluster research for the testing of the basic assumption behind these concepts, which emphasise the role and interaction of the meso-level actors in growth and development[363]. The perceived advantage of actor-centred institutionalism is its ability to explain the interactions between complex actors at the meso-

---

[354] SCHIMANK (2002: 37ff.)

[355] SCHIMANK (2002: 145f.)

[356] SCHIMANK (2002: 146ff.)

[357] WEBER (1922: 1ff.)

[358] SCHIMANK (2002: 36)

[359] SCHIMANK (2002: 88ff.)

[360] See the publications by SCHARPF and MAYNTZ

[361] The definition of institutions or institutional structures leans on NORTH'S understanding of institutions. See *chapter 3.1.*

[362] SCHARPF (2000: 17)

[363] See *chapter 3.1*

level, and the interactions between the actors originating from different analytical levels (macro, meso, micro)[364]. It is an approach to social scientific explanation, which combines the perspectives and experiences from actor and institution research with the rational choice approaches, which are common amongst the economic sciences[365]. *Figure* 5 provides a schematic overview of the basic assumptions of actor-centred institutionalism from the political sciences point of view[366].

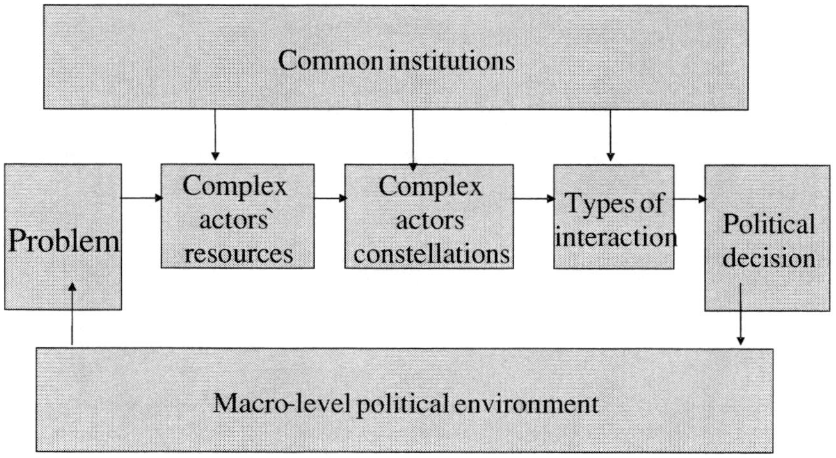

Figure 5. Factors influencing complex actors` political decision-making (Transl. and modif. from Scharpf, 2000)

The classification of the actors in *table 9* pays tribute to one important problem of actor-centred analytical approaches in the analysis of systemic processes. Although it is inevitable that only individuals are capable of handling, these individuals frequently act in the name of another individual or groups of individuals[367]. However, the collective of individuals constituting a complex actor, as e.g. an organisation, determines through its interaction and strategic orientation the collective action of the complex actor as a whole. This aspect relates

---

[364] The evolutionary economist Nelson recommended the actor and institution-centred approach to explain economic development, as *"...classical economics was both evolutionary and institutional. With the rise of neo-classical economics, both evolutionary and institutional aspects were squeezed out of main line economic theory"*. NELSON (2002: 17)

[365] Compare SCHMIDT (2003: 318ff.) and CROUCH (2003: 71ff.)

[366] Adopted from SCHARPF (2000: 85)

[367] SCHARPF (2000: 96)

to system and organisation-theory considerations, and is, in the context of this study, assumed to be negligible[368]. To facilitate the applicability of the actor-centred analytical perspective, the non-individual actors are called complex actors and are treated as one individual actor. These complex actors handle collectively and decide on the orientation of each action through internal interactions, which for the application of the actor-centred research approach are not of empirical interest and can be assumed[369].

Table 9. Classification and basic features of actors[370]

	Aggregated Actors	**Complex Actors**				**Corporative Actors**
		Coalition	***Club***	***Social movement***	***Association***	
Execution of action	Individual	Collective	Collective	Collective	Collective	Organisation
Aim of action	Individual	Individual	Individual	Common	Common	Organisation
Control over resources	Individual	Individual	Common	Individual	Common	Organisation
Decision-making	Individual	Agreement	Poll	Consensus	Poll	Hierarchical
Barriers to entry and exit	Individual capacity	Mutual understanding	Conditional and resource bound entry. Low exit barriers	Voluntary, large number of members, usually no barriers	Sometimes obligatory, usually bottom-up organisation	High barriers, top-down organisation,
Example	Parallel handling, but not organised enterprises	Permanent and voluntary arrangement between firms to extend opening hours	OECD; EU; NGOs	People's movement for environmental protection	Agricultural producer associations	Ministries

The individual and the complex actors interact with each other in one way or another: they may compete with each other in the market, hold producer-client relationships, provide credit or offer educational and extension classes, enforce and obey laws, struggle for the benefits of economic promotion strategies or conduct research with one another. When two or more of these actors interact, this process creates actor constellations and the interactions follow

[368] For detailed information on social systems, society, actors and interactions from a system-theoretical perspective see LUHMANN 1994.

[369] SCHARPF (2000: 97f.)

[370] Adopted from SCHARPF (2000: 105), modified.

certain patterns. These procedures and patterns constitute part of the institutions understood by DOUGLAS NORTH and the '*New Institutional Economics*'[371]. According to SCHIMANK, these institutions constitute structures of societal life, which are subject to permanent and dynamic negotiation of the actors. SCHARPF declared that the institutions are one of the most important resources the actors are endowed with in the pursuit of their aims[372]. Within and through these institutions, the actors act and interact strategically to pursue their aims.

The capability to act strategically requires the intentional orientation of the action. The action should 1) refer to one of the actors (as described in *table* 9), 2) be based upon the actor's cognitive awareness and interpretation and 3) take into consideration the actor's preferences, such as self-interest, compliance with the normative role and the creation and maintenance of a certain identity[373]. Actors initiate strategic action in order to resolve problems and conflicts, and to pursue their perceived optimal goals. Strategic handling is restricted by the limited knowledge of the actors and their resources[374].

HOOD investigated complex actors with regard to their internal structures, and how they affect their external interaction[375]. These behavioural patterns were described as: 1) fatalistic, which means that there is no great interest in co-operation, nor is there much interest in representing ones own position. SCHARPF stated that such actors have no strategic capabilities; 2) individualist, which describes the typical market and bargaining behaviour of the actors. Co-operation and interaction with such actors is common, but usually in the form of 'marriages of convenience'[376]; 3) egalitarian, which is interaction with a high degree of mutual participation and optimal decision-making processes with a minimum of institutions; and 4) hierarchic, which represents a high degree of interaction and co-operation, strongly regulated by institutions and top-down decision-making. HOOD developed a classification of the behavioural patterns of such complex actors, which is shown in *table 10.*

[371] See *chapter 3.1*

[372] SCHARPF (2000: 76ff.)

[373] SCHARPF (2000: 110ff.); SCHIMANK described an actor model (not widely recognised), the major driving force for action of which was the 'sustaining of identity' (SCHIMANK, 2002: 121ff.).

[374] SCHARPF (2000: 107ff.)

[375] HOOD (1998)

[376] BIGGS AND SMITH (2003: 1751)

Table 10. Four patterns of complex actors` strategic orientation[377]

	Complex Actor Cohesiveness	
	Low	*High*
Strong institutional setting	***Fatalist*** Low cooperation; Lack of commitment to collective action; Institutions are limiting; Lack of planning, initiative and foresight; Inertia and passivity	***Hierarchical*** High cooperation; High predetermined set of institutions; Trust in authority and expertise; Unwillingness to make decisions at lower levels of authority
Weak institutional setting	***Individualist*** Market solutions preferred; Rivalry and competition encouraged; Co-operation stresses bargaining; Corruption, extortion and bribery possible;	***Egalitarian*** High participation in decision-making; Favours advocacy and empowerment processes; Too many conflicts inside the organisation; Unwillingness to accept higher authorities

The ability to act strategically is inherent to the actor-centred institutionalism approach. The actors draw on their resources in the pursuit of their aims[378]. These actor-related resources are: personal or organisational characteristics, such as intelligence, strength, and human capital resources as described by COLEMAN (1990); the material resources, including money, land, military power, technological resources, access to information; and the institutional context, which allocates competences, authorities and the rights to participate in decision-making, as well as access to the aforementioned material and personal resources.

The basic actor models and their respective motivations for interaction result in patterns of interaction, which were described by SCHARPF, and are summarised in *table 11*. Actors may interact according to network patterns, with the premise that interaction is geared towards bidirectional negotiations and the emancipation of the partners. Others interact with actors who are higher up in the hierarchy, and depend on the interaction form dictated by their partners. SCHARPF examined a variety of actor-constellations within different institutional contexts, applying explanatory approaches from game theories[379]. He described the forms of interaction as the manifestation of the actors' strategic orientation and capacity, and as dynamic under changing institutional contexts[380].

[377] Compiled according to HOOD (1998: 9) and BIGGS and SMITH (2003: 1752)

[378] SCHARPF (2000: 86)

[379] See SCHARPF (2000: 167ff. and 229ff.).

[380] SCHARPF (2000: 92ff.)

Table 11. Interaction forms of actors in different institutional interaction contexts[381]

Form of interaction	Institutional context			
	Anarchy and minimal institutions	*Networks, regimes and systems of forced negotiation*	*Associations and representative assemblies*	*Hierarchical organisations and the state*
Unidirectional handling	X	X	X	X
Negotiation	(X)	X	X	X
Majority decisions	-	-	X	X
Hierarchical steering	-	-	-	X

3.4 Problem Statements and Research Hypotheses

The theoretical discussion within this study has highlighted the deficiencies in the theories explaining the driving forces behind technological progress. It was revealed that technological progress is a result of actor interaction, and that this interaction is strongly influenced by factors beyond the scope of economic theories. The conclusion derived from the theoretical discussion is that there are relationships between growth and development at the level of inter-actor interactions, which are of crucial importance to the innovation process. These interactions have only been poorly examined for the forest resource-based sectors in developing countries, although these sectors constitute an important planning parameter in the development strategies of rural areas. *Figure* 6 brings together the various theoretical strands discussed in relation to this study.

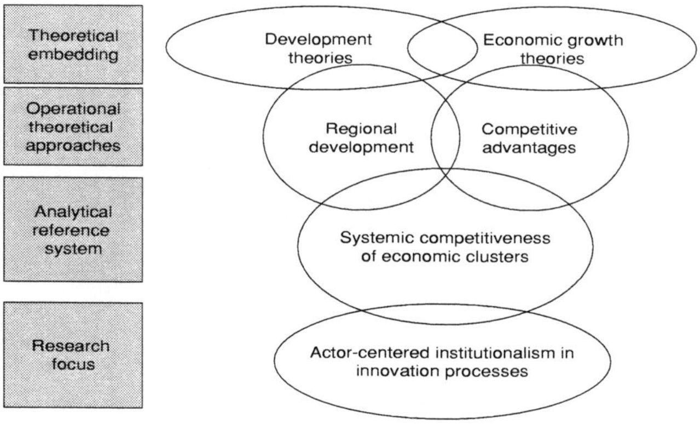

Figure 6. Theoretical strands relevant to this study

[381] Translated and modified from SCHARPF (2000: 91)

Problem Statements

The theoretical discussion revealed the following statements regarding the relationships between economic growth, development and forest resource-based sectors in developing countries:

- Economic growth is essential to development. However, growth is only one aspect of sustainable development.
- Technological progress has been identified as one major driving force behind competition and economic growth. However, the factors underlying this process have not yet been fully explained by the various strands of economic growth theories.
- In natural resource-based economic clusters, the value-adding processes based on technological progress are essential for the employment of these resources in the national and regional economies, and their integration into development strategies. Value-adding processes are scarce in the forest-based sectors of developing countries.
- The factor underlying technological progress, the innovation process, originates from entrepreneurial behaviour and assets, as well as from social behaviour and assets, i.e. human capital and the interaction between actors. The innovation process is a socio-technical system, which depends on various parameters of development and the social interplay between the actors
- Innovation research has hardly been conducted in developing countries. Where it has taken place, macro-economic approaches were applied principally. However, there is a need to investigate innovation processes at a systemic level.
- To investigate innovation processes at a systemic level, actor-centred approaches under given institutional aspects are suitable, although such approaches have not yet been common.
- The role of the meso-level actors in the innovation process in developing countries, their strategic capabilities, and the interactions between the meso and the micro-level, has not yet been investigated sufficiently.

Guiding Hypotheses

Subsequent to the problem statements, the following research guiding hypotheses were formulated for this study:

- Innovation is a result of interactions at the meso and the micro-level, as well as the interactions between these levels, as distinguished in the concept of systemic competitiveness.
- The ability of meso-level actors to enhance the innovative capacity of the micro-level entrepreneurs depends on their own strategic capability.
- The actor constellations and interactions within the meso-level influence their strategies for the promotion of the micro-level.
- Technological progress (innovation) at the micro-level follows a path/trajectory, and is driven by intra-level interaction (diffusion).
- The intra-micro-level and the intra-meso-level interactions are influencing the effectiveness of the promotion of innovations.

4. Background of the Study

This chapter provides an introduction to the study area and illustrates the natural, social and economic situation, which is important for the understanding of the interrelations between the regional bamboo sector and the social and economic components in the context of this study. It begins with the justification of the choice of the study region and continues with the description of the basic socio-economic and natural features.

4.1 Selection of the Study Region

As the theoretical discussion on the role of actors and interactions in innovation and growth processes has shown, most of the research in this area has been carried out in the industrialised countries of Europe and North America. However, such research, especially in the developing countries, has become more and more urgent as the integration of these regions into the increasingly globalised world market challenges their economic sectors to increase their rates of innovation and technological advance (see *chapter 2*).

Most developing countries are still suffering from the worldwide structural crisis of the 1990s. They have adopted strategies to overcome the negative consequences of this crisis following recommendations made on the basis of the world community's economic expertise and guidance[382]. In many regions of Latin America projects, promotion activities and adjustment programmes have been established in recent years, all with the aim of supporting local and regional economic sectors. The Latin American agricultural sectors in particular, which were heavily hit by the crisis, are steadily undergoing new political and economic adjustment initiatives[383]. One important part of these strategies is the initiation of technological advance and innovation processes through the transfer of knowledge and technology, by means of foreign direct investment and the formation of national and regional innovation systems (see *chapter 3*).

The Colombian coffee growing region represents one case of a Latin American region, which has been affected by the continental and worldwide structural and agricultural crisis. There is great interest on the part of regional, as well as national actors to counteract the ongoing process of economic decline of traditional regional structures. In the difficult social and political context of Colombia, the '*Eje Cafetero*' was the most stable and secure Colombian region for many years. Economic stability and prosperity based on the burgeoning coffee sector, as well as the low rates of unemployment contributed to this positive regional

[382] I.e. stabilisation programmes supported by the World Bank and the International Monetary Fund (SCOBIE AND JACOBSEN, (1995: 3ff.)

[383] TABOR (1995: 23ff.)

climate[384]. Since the beginning of the 1990s the region has become increasingly subject to the activities of the armed factions, who attack state authority, plant illicit drugs[385] and engage in other violent acts. This goes hand in hand with the continuing fall of world market prices for coffee and the general structural crisis of the agricultural sectors in Colombia, the region has become a common harbour for the refugees from Colombia's most violent areas. These refugees arrive in a region already greatly weakened, without possibility of employment and almost no social welfare system in place to care for them (see *table 12*).

Currently, the Colombian actors are trying to stabilise the region and provide alternative sources of income for the native population, as well as for the recently arrived refugees. In addition to a number of agricultural products, such as organic coffee production, bananas, citric fruits, flowers and avocados, the region's natural resources are considered to constitute part of the economic and social development process[386]. One of these natural resources is the native bamboo species *Guadua angustifolia*, of which there is about 30 000 ha in the study region (see *table 15*)[387]. Apart from the fact that the bamboo stands are virtually the only natural resource accessible to private small-holders, the regional bamboo sector is also an established part of the regional economy, albeit not a very productive one (see *chapter 4.4*).

The public and private actors in Colombia have increased their activities in order to promote the use and the industrialisation processes of the regional bamboo production and processing sector. Furthermore, a number of international co-operation and research projects have been initiated since 1999. This year represents the turning point in terms of the regional recognition of the bamboo resources in the Colombian coffee region. In January 1999 an earthquake destroyed the capital city (Armenia) of the department of Quindio and a number of smaller cities and villages in the region. The traditionally constructed buildings made of bamboo, timber and mortar resisted the quakes, whereas modern buildings collapsed and left several hundred people dead in the affected areas. Reconstruction programmes initiated by the German technical cooperation agency (GTZ) and the national Colombian fund for the reconstruction of the coffee region (FOREC[388]) resulted in the rebuilding of three villages with bamboo constructions and a number of bamboo housing social projects were instigated. Funds were also provided to encourage reforestation with Guadua in the coffee region, in order to enlarge the resource base for further housing projects.

---

[384] The economic power of the agricultural and especially the coffee sector in the Colombian national economy is based upon a system of unequal distribution of land and capital assets, which paved the way for the armed political conflict in the 1950s and 60s (DILGER, 1996: 49ff.; GTZ 2001b: 26f. and MOSER, 2000: 9f.). For a detailed discussion of the Colombian land tenure and distributional conflicts see MACHADO,1998.

[385] The area of Colombia under illicit drug plantations has more than doubled between 1991 and 1999. Today more than 110 000 ha are dedicated to the cultivation of such plants (GTZ, 2001b; 26ff.).

[386] CARRIZOSA (1992: 163ff.) and VILLARONGA-WALKER (1998)

[387] In the further course of the present work the terms bamboo and Guadua are used synonymously.

[388] FOREC = *Fondo Reconstrucción del Eje Cafetero*

4.2 Basic socio-economic Information on the Study Region

The study region consists of the departments[389] Caldas, Risaralda, Quindio and the Valle de Cauca[390]. These four departments form the Colombian Andean coffee growing region known as '*Eje Cafetero*[391]'. The region is located in the inner valley of the Colombian Andes at altitudes of between 1 000 and 1 800 m ('*Tierra templada*'), which constitutes the optimal tropical altitude for the cultivation of coffee. *Table 12* reveals the basic socio-demographic features of the four study departments. Maps of the study region can be found in the *annex 1*.

Up until the year 1966 the three departments Caldas, Quindio and Risaralda formed a single department named Caldas (today known as '*Viejo Caldas*[392]'). Due to political differences in the previous government and increasing competition between the growing urban centres of Manizales, Armenia and Pereira, the original department split up. The newly created departments are amongst the smallest in Colombia. As the economic situation of the state and that of the departments of the coffee region requires extensive public funds, there are plans to reunite the three departments, or at least unite some administrative entities, such as the regional environmental authorities (the CARs).[393]

Colombia's national economy has undergone years of minimal growth, stagnation and recession since the beginning of the 1990s. Colombia has performed better than most of the neighbouring Latin American countries, however, and was able to stabilise its economy by the end of the 1990s, whereas countries like Ecuador and Argentina had to 'dollarize' their economy and still suffer from the consequences of inflation and recession[394]. Colombia was able to slow inflation down[395] and to recover from decreasing and negative GDP-growth rates. In the national economy, the departments of the study region play a stabilising role. Their rates of unemployment are below the national average of 20%[396] and their contribution to the GDP is situated in the medium to lower ranks[397].

---

[389] The term *departamento* refers to the hierarchical mid-level administrative unit between the state government and the municipalities. In Colombia 32 of these administrative divisions constitute a federal-like system. In the present study *departamento* is translated as department according to the translation applied by the CIA fact book (www.cia.gov/cia/publications/factbook/).

[390] Only the northern part of the department of the Valle de Cauca officially belongs to the coffee growing region. However, in the southern regions of this department bamboo is a common plant and is widely grown on the sugarcane plantations as windbreaks, and for watershed and soil protection purposes. In these parts of the department the sugarcane producers and processors make up the group of actors interested in the resource.

[391] Transl.: Coffee axis

[392] Transl.: Old Caldas

[393] CINEP (1998: 39ff.)

[394] Ecuador introduced the US$ in the year 2000 and in Argentina a 1:1 dollar parity was adopted temporarily in the year 2001.

[395] The inflation rate could be slowed from 25% in 1992 down to less than 8% in 2001 (Dresdner Bank Database enquiry 2004)

[396] REPUBLICA DE COLOMBIA. DEPARTMENTO ADMINISTRATIVO NACIONAL DE ESTADISTICA (DANE). Internet database enquiry 2004.

[397] Note: The three study departments Caldas, Quindio and Risaralda belong to the smallest of the 32 Colombian departments.

Table 12. Socio-demographic and economic features of the study departments, 2001[398]

Depart-ment	Capital city	Area in km²	Popu-lation	Hosted inland refugees[399]	% of national GDP 2001	Un-employ-ment (%)	Sub-employ-ment (%)
Caldas	Manizales	7 888	920 000	15 000	2.2	14.1	41.4
Quindio	Armenia	1 845	430 000	6 000	0.9	17.7	31.9
Risaralda	Pereira	4 140	750 000	14 000	1.7	17.0	33.9
Valle de Cauca	Cali	22 140	3330 000	56 000	11.4	17.4	35.6
Total	-	36 013	5430000	91 000	16.2	-	-

As already mentioned, the study region was and remains to this day the most secure region in Colombia. The situation is tense and fragile, however, especially in the departmental areas at the slopes of the Andes and close to the neighbouring regions. The increasing insecurity in the country side has led to increased migration into the region's urban centres. The coffee crisis and the migration flows result in increased unemployment, which in turn feeds economic and political violence[400]. THE WORLD BANK has published survey results from the year 2000, in which the Colombians were asked to define the major problems and causes for the poverty of their state. In this survey the four most highly ranked problems were violence, unemployment, education deficits and corruption[401]. In addition to the need for the reform of state institutions and organisations, and the initiation of counteracting programmes and measures[402], the negative role of violence and crime in economic growth and poverty was identified as an important factor in Colombia, which has to be resolved in order to improve the overall situation of the country[403].

The basic assumption behind the countermeasures initiated and planned by the Colombian state is that the generation of increased economic growth and the equal distribution of the surplus value will reduce poverty and this in turn will contribute significantly to the reduction of crime and violence in the country[404]. In the further context of this study economic violence will be referred to as 'crime', whereas politically motivated violence will be termed

[398] REPUBLICA DE COLOMBIA. DEPARTMENTO ADMINISTRATIVO NACIONAL DE ESTADISTICA (DANE). Internet database enquiry 2004.

[399] Figures according to the official listings of the Colombian government of 2003. The total number of inland refugees in Colombia is about 1 000 000 people (REPUBLICA DE COLOMBIA, MINISTERIO DE PROTECCIÓN SOCIAL, 2003).

[400] MOSER (2000: 12ff.) distinguishes three types of violence: political, the economic and the social violence.

[401] THE WORLD BANK (2002: 85)

[402] See FUHR (1998)

[403] THE WORLD BANK (2002: 31ff.). For the role and costs of violence and conflict in other countries of the world see THE WORLD BANK (2003).

[404] GTZ (2001b: 26ff.);

'violence'. In Colombia, however, all types of violence are closely interrelated, which makes it difficult to identify and distinguish clearly the causes and effects of the individual types of violence. The attempts of the Colombian government to counteract the worsening economic and social situation include military action against armed factions and the fumigation of illicit drug production sites[405], as well as social and economic support programmes[406].

4.3 The agricultural Sector of the Study Region

The Colombian agricultural sectors contribute 19% to the Colombian GDP and provide the second most important export commodities behind fossil fuel energy sources, such as petroleum and coal[407]. In total, 30% of the Colombian labour force is occupied in the agricultural sectors. Since the late 1980s, however, these figures have declined continuously. Between 1995 and the year 2000, the number of unemployed in rural areas has almost doubled to 10%[408]. In the same time period the value of the exported agricultural products (incl. coffee) has fallen almost 25%[409].

In Columbia, as in almost every Latin American country, the agricultural sector is suffering from the crisis of the national economy. Major aspects of this crisis are the globalisation of markets for agricultural commodities, increasing competition and decreasing producer prices[410]. Internally, these countries face problems in adjusting the agricultural sectors to the new situation due to land tenure conflicts, lobbying of national elites and the intervention of multi-national actors[411]. Public policies appear to be helpless in terms of meeting national and international challenges, especially in the Colombian context, where the state's legitimacy is called into question by the ongoing conflicts[412]. These challenges are encountered in Colombia through the continuation of the peace process, the reform of the agrarian sector and the fight against the clientelist structures in the Colombian society[413].

[405] Military action against the illicit drug production has been supported by the United States of America in the form of 'Plan Colombia', which was introduced by the Clinton administration and the Colombian president Pastrana in the year 2000 (with a budget of US$ 664 million over five years for the armed fight against the drug production). These activities are to be flanked by financial support measures promoting social programmes with a budget of only US$ 81 million for the same timeframe. (GTZ, 2001b: 30f.)

[406] Besides the loss of human life and political stability there is also an economic dimension to the violence in Colombia, which is expressed in decreasing GDP and GNP growth rates and decreasing rates of investment in the country (MOSER, 2000: 25ff.).

[407] REPUBLICA DE COLOMBIA. DEPARTMENTO ADMINISTRATIVO NACIONAL DE ESTADISTICA (DANE). Internet database enquiery 2004

[408] REVISTA NACIONAL DE AGRICULUTRA (2001b: 69)

[409] REVISTA NACIONAL DE AGRICULUTRA (2001b: 70)

[410] MACHADO (1998: 15ff.)

[411] MACHADO (2002: 67ff.) and HAGEN (1997: 7ff.)

[412] MACHADO (1998: 31ff.)

[413] KALMANOVITZ (2001a: 39ff.)

For many decades the coffee sector constituted the major source of income in the study region[414]. All over Colombia the coffee sector provides 35% of the employment generated in the national agricultural sector, which means that more than 500 000 families depend on coffee for their income[415]. Since the early 1990s the Colombian coffee sector has faced a severe crisis caused by decreasing world market prices and new competitors from Asia[416]/[417].

The sector has not yet fully recovered from the worst years of crisis in the mid '90s[418]. Between 1980 and 2000 Colombia lost about 20% of the 1 000 000 ha under coffee cultivation, which existed in 1980[419]. However, this still accounts for 24% of the total agricultural area cultivated by the Colombian farmers[420]. An area of almost 300 000 ha is under coffee cultivation in the study region, with more than 100 000 farms (*'Fincas'*) dependent on the production of coffee. The majority of these production units are based upon a household production system in which the family members provide the labour force and capital inputs[421]. *Table 13* highlights the most important figures in relation to the coffee sector. Other important agricultural sectors of the region are cattle farming, banana and flowers, which have gained importance in recent years. These products are mainly cultivated by medium and large scale farmers, who possess sufficient arable land resources to diversify and amplify their agricultural production.

The Colombian coffee sector is organised as an agricultural productive chain[422], as is common in the industrial and agro-industrial sectors and branches of many Latin America countries[423]. The most important organs of the Colombian coffee sector are the coffee growers' federation FEDECAFE[424] and the research entity CENICAFE[425]. The federation is based upon a democratic hierarchy and officially works independently from state policies and funds. However, the state closely co-operates with the organisation, as FEDECAFE is responsible for the provision of public services to the coffee farmers, such as health care, installation of education facilities and the provision of infrastructure. Probably the most

---

[414] For a detailed description of the coffee culture and colonisation in the region see FORERO (1999: 37ff.).

[415] SUAREZ (2001: 38f.)

[416] Vietnam has replaced Colombia as the second largest coffee producing country behind Brazil (PIZANO, 2001: 57ff.).

[417] This is not the first crisis to hit the Latin American coffee sector. Between 1920 and 1936, the coffee sectors of Latin America suffered from changing world market structures and the overall impact of the 1929 economic depression (SAMPER-KUTSCHBACH, 1995).

[418] The producer prices for coffee in 1999 were below the prices in 1991 (SUAREZ, 2001: 89).

[419] PIZANO (2001: 9)

[420] SUAREZ (2001: 40)

[421] FORERO (1999: 197ff.) and FORERO ET AL. (2002: 116ff.)

[422] *Cadena productiva agricola* or *cadena productiva sectoral*

[423] ESPINAL AND ROLDÁN (2000: 18f.)

[424] FEDECAFE = *Federación Nacional de Cafeteros de Colombia*

[425] CENICAFE = *Centro Nacional de Investigaciones de Café*

important activities carried out by the organisation are the co-ordination of a common marketing strategy and the international export of coffee.[426]

Table 13. Important figures of the coffee sector in the four study departments (1997)[427]

Depart-ment	Total area under coffee production (ha)	Average coffee area/finca (ha)[428]	No. coffee producing farms	Percentage of coffee growing households living in misery[429]	Participation of coffee in departmental GDP in %[430] in 2001
Caldas	91 000	1.9	47 000	10	9.6
Quindio	51 000	6.4	8 000	6	4.3
Risaralda	62 000	2.7	23 000	10	3.9
Valle de Cauca	90 000	3.3	27 000	11	1.2

In addition to the powerful coffee producing sector, there are several other agricultural sectors in Colombia, which basically work along the same lines. As the topic of the productive chain will be raised again in the further course of this study, the basic concept underlying this system is briefly introduced in the following paragraphs.

The large agricultural sectors in Colombia: the coffee sector, the sugar sector and the cattle farming sector are organised in productive chains, which are less dependent on state support and subsidies than the chains of the smaller agricultural sectors. This is due to the large number of members and the financing power of these chains. These sectors are not included in the national programmes for the organisation and promotion of productive chains. They work according to the same basic features, however.

No common definition could be found for the Spanish term '*Cadena productiva*' ('productive chain' in English). In some contexts the term is used synonymously with 'cluster', as described in *chapter 2.2*. In other contexts the term is used to describe the value-chain of raw material production and processing sectors. Basically, the productive chain is a meso-level concept, which makes possible adherence to and influence over the path of production, processing and marketing of a product (product group)[431].

---

[426] SCHNEIDER (1998: 11ff.) and BENTLEY AND BAKER (2000)

[427] Figures as published by BENTLEY AND BAKER, 2000

[428] The national average land area of the more than 420 000 Colombian coffee growing farms was 1.5 ha in 1997 (BENTLEY AND BAKER, 2000: 1). Only 5% of the coffee growing fincas throughout Columbia cultivate more than 5 ha of coffee (REVISTA NACIONAL DE AGRICULUTRA, 2001a: 15).

[429] The Colombian coffee growers' federation (FEDECAFE) defines misery as comprising two or more of the following: poverty, crowding, inadequate housing, inadequate access to drinking water, children not attending school (BENTLEY AND BAKER 2000: 7).

[430] REPUBLICA DE COLOMBIA. DEPARTMENTO ADMINISTRATIVO NACIONAL DE ESTADISTICA (DANE). Internet database enquiry 2004

[431] MACHADO (2002: 237f.)

In Colombian agricultural policy and planning of the '*cadenas productivas*'[432]/ [433], of which there were ten in Colombia up to 2002, refers to the agreements between public and private actors from a specific sector on the common strategic orientation in production, processing and commercialisation, and the improvement of international competitiveness ('*Acuerdos de competitividad*')[434]. These chains were: the aviculture and piggery chain, cotton and textiles, oil seeds, the potato chain, the cacao chain, the horticulture chain, the forestry chain, the chain for dairy products, the chain for the cultivation of shrimps, and the "*panela*[435]" chain[436]. Usually each productive chain is endowed with a research entity, which is co-financed by both chain and public funds[437]. *Table 14* shows examples of some of these Colombian chains.

Table 14. Examples of Colombian 'Cadenas Productivas', their producer organisations and research entities[438]

Productive chain	Acronyms for the sector's organisations	Acronyms for the related research entity
Coffee	FEDECAFE	CENICAFE
Sugar	ASOCAÑA/FEDECAÑA	CENICÑA
Oil palm	FEDEPALMA	CENIPALMA
Timber and wood products[439]	In formation	CONIF[440]

[432] ESPINAL AND ROLDÁN (2000: 38ff.) and CHICA (1996: 11ff.)

[433] In Colombia these productive chains are legitimised by the Colombian Agricultural laws No. 101 of 1993 and No. 811 of 2003.

[434] This concept was initiated by the Colombian government through the creation of the '*Consejo Nacional de Competitividad*' in 1994 and from then on continuously extended and developed (HERRERA, 2000: 38ff. and ESPINAL AND ROLDÁN (2000: 11ff.)

[435] 'Panela' is made of sugar cane.

[436] Apart from the named productive chains there were several other competitive agreements in other sectors in the year 2002, namely the cultivation of citric fruit, banana production, the productive sectors for rice and for tobacco. However, these agreements had not yet reached the state of an officially recognised productive chain. REBUBLICA DE COLOMBIA. MINISTERIO DE AGRICULTURA (2002: 61f.)

[437] The legal support and governing framework of the agricultural sector in Colombia is described in detail in REPUBLICA DE COLOMBIA. CONTRALORÍA (2002).

[438] SUAREZ (2001) compares the Colombian agricultural sectors and their forms of organisation and co-ordination with each other. She describes the advantages and disadvantages of the sectors and their organisation forms.

[439] In the timber producing and wood processing sector only competitive agreements have been established in the past. Until 2004 the forestry productive chain was dominated by the large scale reforesters and the pulp and paper industry. A co-ordinated and organised chain is not yet in place (see *chapter* 6.4.3).

[440] CONIF was founded as a state research institute, which shall become a private-public institute when the forestry productive chain is established.

4.4 Forest and Bamboo Resources in the Study Region

The current state of the natural environment in the study region is closely linked to its history of settlement and coffee cultivation. The Spanish colonisation of the region started in the late 18th century with the foundation of settlements in an area, which today forms part of the department of Caldas[441]. Before the Spanish entered this area, however, indigenous tribes, such as the *Quimbaya* or the *Pijao*, had settled and cultivated the region. The botanist Alexander von Humboldt, who visited Colombia in the early 19th century, described the indigenous cultures of that time, the vast tropical forest areas, the Andean mountain and volcano vegetation and the coastal regions of the country. In 1801 he travelled through the valley of the Cauca river and other parts of today's coffee growing region, where he described for the first time the large stands of the giant bamboo *Guadua angustifolia* Kunth[442], which occurred as almost natural monocultures, and in the understorey of the still abundant natural forests[443].

The role of this natural bamboo resource in pre-Colombian times and during the history of settlement and cultivation of the study region has been mentioned and emphasised by a number of regional scientists and authors[444]. '*Guadua y Café*[445]' has become a set term in the description of the region's typical cultural and natural characteristics. Bamboo and bamboo products have been applied and partially replaced timber products in the region, especially in the construction of houses and for auxiliary uses in the agricultural production systems (the '*fincas*'). During the colonisation of the region almost all natural forests and large areas of bamboo stands were transformed into arable land[446]. Last residues of the natural forests have survived where agricultural cultivation was not considered economically feasible, and at altitudes above the optimum growth conditions for coffee. In contrast to the forest resources, the bamboo resources have survived and were even actively maintained and managed in the midst of the cultivated land. Today between 30 000 ha[447] and 40 000 ha can still be found in Colombia; about 50% of them in the study region (see *table 15*)[448].

---

441 The colonisation process of the region was named the '*Conquista Antioqueña*', as most of the settlers originated from the region known as Antioquia (capital Medellin). These settlers brought the Antioquenian traditions and culture, known as the '*Cultura Paisa*' in Colombia, to the study region (CINEP, 1998: 15ff.).

442 Humboldt named the plant '*Bambusa Guadua*'. The exact botanical identification and taxonomical description was subsequently carried out by the German botanist Karl Sigesmund Kunth in the year 1822.

443 A. V. HUMBOLDT (1821): Personal narratives of travels to the equinoctial regions of the new continent during the years of 1799-1804. Longman UK

444 HIDALGO (2003: 366ff.); MORAN (2001); LONDOÑO (1992)

445 Transl.: 'Guadua and Coffee'

446 BERNAL (2002: 58f.). According to latest site classifications more than 40 000 km² in the MSBC-project pilot region are suitable for *Guadua angustifolia* plantations, which includes the Eje Cafetero plus the department of Tolima (total area of the pilot region: 58 000 km²) (BERNAL, 2002: 10f.). This indicates that before the colonisation this area may have been stocked with Guadua and mixed Guadua-forest stands.

447 GIRALDO AND SOBOGAL (1999: 35)

448 For a detailed description of the current distribution of forest and bamboo resources in the study region see BERNAL (2002)

Table 15. Bamboo resources in the study area[449]

Department	Area *Guadua angustifolia* (in ha)
Caldas	2 019
Quindio	5 215
Risaralda	2 972
Valle de Cauca	7 520
Total	**17 726**

The bamboo stands represent almost the only natural resource in the study region accessible to and utilisable by private entrepreneurs and small-holders. Practically all the stands are located on private property. The legal administration and control of these stands lies in the hands of the regional environmental authorities, the CARs[450], which represent the regional branches of the Ministry of the Environment. In the study region almost 3 000 ha of Guadua have been actively planted since the early 1980s[451]. These planted bamboo stands are not officially the CARs' responsibility, but are under the control of the representative regional offices of the Ministry of Agriculture and Regional Development[452]. In Colombia forest resources planted for commercial purposes are integrated into the national agricultural policies, and in the broader context into the agricultural productive chains[453]. According to an agreement between the two ministries, however, the administration and control of all bamboo resources in the study region is carried out by the Ministry of the Environment and the CARs.

The CARs are the legitimate state organs put in place to administer and control the natural resources in the departments of Colombia. They draw their legitimacy from the Colombian environmental law[454]. Each department is endowed with one of these entities (see *table 16* for the list of CARs in the study region), which is authorised to execute autonomously the national policies, plans and programmes in relation to environmental issues. One important function of the CARs is the assignment of concessions and permits for the use of natural resources. In line with this function, the CARs of the study region, under the guidance of the national programme for the sustainable management of forests in Colombia (MSBC[455]),

[449] Figures as published by BERNAL in 2002

[450] CAR = *Corporación Autonoma Regional*

[451] CASTAÑO AND MORENO (2004: 40)

[452] REPUBLICA DE COLOMBIA. MINISTERIO DEL MEDIO AMBIENTE (1999: 513) and as defined by the Colombian Law No. 101 of 1993 (*Ley general de Desarrollo agropecuario y pesquero*) on the formation of productive chains in the agricultural sectors (supplemented by the Colombian law No. 811 of 2003).

[453] For a detailed discussion of the Colombian Forestry Productive Chain see *chapter 6.4.3*.

[454] *Ley del Medio Ambiente*. Ley No. 99 de1993 Art. 31

[455] MSBC = *Manejo Sostenible de Bosques en Colombia*. This project was initiated by the German technical co-operation and the Colombian Ministry of the Environment in the year 2000. In 2002 there were two pilot areas of the project, one of them located in the *Eje Cafetero*, plus the department of Tolima.

introduced a common regulation on the use of the bamboo resources in the area of their responsibility[456].

Table 16. List of CARs in the study region

Department	Acronym for the departmental CAR
Caldas	CORPOCALDAS
Quindio	CRQ
Risaralda	CARDER
Valle de Cauca	CVC

In addition to the regulation of management and utilisation of the resource, the CARs have developed a promotion approach for the Guadua sector[457] in co-operation with other regional meso-level actors. In the official publications and public appearances, such as websites and other dissemination material, the CARs termed this approach the 'Guadua Cluster'[458], whereby, the term cluster refers rather to the translation of the '*Cadena productiva*' promotion and organisation concept[459]. The organisation of the Guadua cluster and the related actors in 2002 was still in its infancy and hardly more than a concept on paper, with some non-binding agreements between meso-level actors.

4.4.1 *Guadua angustifolia.* Uses and Use Regulations

For a better understanding of the remainder of this study some basic characteristics of the species *Guadua angustifolia* Kunth are described in the following paragraphs.

Latin America hosts more than 25% of the bamboo species existing worldwide[460]. One of these species is *Guadua angustifolia* Kunth, one of the woody 'giant' bamboos of the Americas. The natural distribution of the genus *Guadua* covers almost all of South and Central America. Members of the genus *Guadua* constitute the most extensive bamboo forest in Latin America, covering more than 120 000 ha[461]. *Guadua angustifolia* can be found naturally from Panama to Ecuador, northern Brazil and Bolivia[462]. It was taxonomically described first by Alexander von Humboldt in the year 1801. These bamboos are perennial

[456] *Norma unificada en Guadua* (MSBC, 2001 and MSBC, 2003)

[457] In the further course of this study, the terms bamboo and Guadua in the Colombian context are used synonymously. The Guadua sector is the Colombian bamboo sector and Guadua based enterprises are enterprises producing and processing the bamboo species *Guadua angustifolia.*

[458] E.g. the CORPOCALDAS brochure: 'Microcluster de la Guadua' of 2002

[459] See *chapter* 2.1 and 2.2 on industrial clusters and *chapter* 4.3 on the '*Cadenas productivas*'.

[460] Currently there are more than 1 250 known species of bamboo belonging to 75 genera (LIESE, 1998: 12).

[461] JUDZIEWICZ ET AL. (1998: 246)

[462] JUDZIEWICZ ET AL. (1998: 239ff.)

and are endowed with a pachymorph (sympodial)[463], amphimorph (amphipodial) or leptomorph (monopodial) rhizome system. *Guadua angustifolia* is a pachymorph species, although its physiognomic appearance is more that of a leptomorph.

The optimum natural growth conditions for Guadua are between 900 and 1 600 m a.s.l. in the tropics, with average annual precipitation between 1 800 and 2 500 mm[464]. Hence, it overlaps with the optimal climatic range for coffee cultivation. As with all bamboos, the Guadua's common means of reproduction is vegetative or asexual via the rhizome system, which enables growth rates of up to 30 m height and diameters of 20 cm in six months, with a maximum daily growth in height of up to 30 cm[465]. The annual CO2-fixation rate of 1 ha of *Guadua angustifolia* was measured as nine metric tons, whereby most of the carbon is fixed in the culms[466].

The stand density of Guadua stands varies depending on the stand quality and climate[467]. However, some average figures have been established based on experiments in the Colombian coffee growing region. The average density in the study region is about 2 900 culms per hectare, whereby, 46% of them have diameters of more than 12 cm[468]. *Guadua angustifolia* is one of the bamboo species that flowers as well, gregariously in cycles and sporadically. Unlike other bamboo species that flower gregariously, *Guadua angustifolia* does not die after the flowering phase[469].

The utilisation and processing of Guadua raw material depends on its origin in the culm. The characteristic bamboo growth is that of a cane, similar to the growth of other GRAMINEAE. The culm is separated into the nodes and the internodal parts. The lower part of the culm is thicker than the upper parts and the internodes longer. The fibres towards the outer culm wall are longer and thicker[470]. Therefore, applications and final uses of bamboo products determine the part of the culm from which the original raw material must be selected.

For all common uses of the bamboo raw material it is important that the culm has reached a certain age, which ensures that the lignification process has been completed. The gross density of the mature wood of *Guadua angustifolia* is between 0.6 g/cm³and 0.8 g/cm³ [471]. Chemically its composition is similar to the wood of trees with a high percentage of silica in

[463] Due to the physiognomic appearance of the bamboo stands, which is usually characterised by the distance between the individual culms, the pachymorph bamboos are commonly known as clumping bamboo species, whereas the leptomorphs are termed running bamboos.

[464] GIRALDO AND SABOGAL (1999: 99f.)

[465] JUDZIEWICZ ET AL. (1999: 239ff.); GIRALDO AND SABOGAL (1999: 67ff.) and HIDALGO (2003: 10f.)

[466] 39.5 tons of the total of 54.3 tons fixed annually per ha Guadua (CASTAÑO AND MORENO, 2004: 43f.)

[467] CAMARGO (in prep.)

[468] CASTAÑO AND MORENO (2004: 66)

[469] HIDALGO (2003: 31f.) describes the gregarious flowering of *Guadua angustifolia* in Colombia in the 1970s and mentions that only few individuals died after the cyclic flowering while the majority of the stand survived.

[470] LIESE (1998: 58ff.)

[471] GIRALDO AND SABOGAL (1999: 148)

the outer layers of the culm wall[472]. The process of lignification starts after the first growth phase (which begins after reaching the final height, development of the branches and unfolding of the leaves) of the culm is completed (between 0.5 and 1 year). According to local knowledge and experience[473] the liginification process in *Guadua angustifolia* finishes after four years, and is referred to as the maturing phase of the culm[474].

Figure 7 shows the Guadua plant and its culm classification (commercial parts) according to the traditional uses of Guadua in the study region. In *table 17* the commercial parts of the Guadua and their final permanent and auxiliary uses are explained. This classification is the result of the long-term market demand, i.e. from the construction sector, which is currently the main consumer of Guadua bamboo in the study region[475]. *Graphic 1* reveals the decreasing volumes of Guadua harvested in the two most important Guadua producing departments. HELD AND MANZANO assume that due to the high consumption of Guadua products in the construction sector, the total consumption of Guadua strongly depends on the cyclic behaviour of this sector. Since 1994, Colombia's construction sector has been in decline, which is manifested in the decreasing volume of m² constructed annually. In *graphic* 1 the possible dependency between the cyclical trends in the Colombian construction sector and the volume of harvested Guadua in two of the study departments is shown.

The demand of the construction sector for Guadua is satisfied by traditional intermediaries and service providers, the '*guadueros*'[476]. The *guadueros* either work as independent contractors employing two to three people and travelling around to harvest the Guadua stands in the region, or they are contracted, sometimes even employed, by medium to large scale farmers to harvest and manage their Guadua stands. If the *guaduero* works independently, he[477] buys the standing Guadua at flat prices of between Col. $ 1 000 and 2 000 per Guadua[478]. He and his crew are responsible for the harvesting, classification and transport of the culms[479]. They provide the standard assortment of Guadua products as described in *table 17* and entertain business relationships with regional wholesalers and retailers, which they

[472] Main contents are: 61-71% celluloses and hemicelluloses, 20-30% lignin, 16-21% other chemical compounds and up to 4% silica. GIRALDO AND SABOGAL (1999: 149)

[473] Usually, the age of the culm is determined visually according to the colour of the culm and the plaits on it (HIDALGO, 2003: 16)

[474] No scientific research into the processes taking place during the maturation of the culm of *Guadua angustifolia* has been conducted so far. LIESE (1998: 130ff.) describes the maturation process and the modifications taking place in the bamboo culm for other species and the Imperial College London is currently working on the characterisation of the cell wall growth during the maturation process of *Guadua angustifolia*. GIRALDO AND SABOGAL (1999: 89f.) and CASTAÑO AND MORENO (2004: 67ff.) also describe visual selection criteria for the determination of the culm age and state of maturation.

[475] See HELD AND MANZANO (2004)

[476] As there is no appropriate translation of the term '*guaduero*' this original term will be used throughout the remainder of this study.

[477] Note that until today, no women were found working as guadueros in the study region.

[478] See page xvi for exchange rates.

[479] For a detailed description of the guaduero system and cost-benefit analysis of their work see ARROYAVE, MESA AND LÓPEZ (2000) and HELD AND MANZANO (2004).

supply more or less regularly. A list of average prices for the standard Guadua products is compiled in *table 18*.

Table 17. Commercial parts of the Guadua culm and their final applications[480]

Commercial part of Guadua culm		Most common final applications[481]
Spanish	English	
Esterilla/ Guadua picada[482]	Bamboo mat/open Guadua	Auxiliary: scaffolding; temporary huts; concrete formworks; etc. Permanent: walls; floors; staircases: tilts; etc.
Cepa/Basa	Bottom part[483] (raw material for the '*esterilla*')	Auxiliary: scaffolding; construction standards; fencing; etc. Permanent: perm. columns; beams; laths; laminated bamboo products; furniture; handicraft; etc.
Sobrebasa	Middle part	Auxiliary: scaffolding; fencing; construction standards Permanent: beams; furniture; handicraft
Varillón	Upper part	Auxiliary: standards in banana and agricultural plantations, scaffolding Permanent: handicraft
Caiman	Rhizomes	Handicraft and furniture
Ramas	Branches	Handicraft

Table 18. Average wholesale and retail prices for standard Guadua products in the study region in 2002[484]

Product	Price in Col. $/meter or piece[485]
Esterilla	1 300/piece
Cepa	400/m
Sobrebasa	280/m
Varillón	350/m
Caiman	varies
Ramas	varies

480 The information in this table was compiled according to the data available in HELD AND MANZANO, 2004.

481 Understood as auxiliary applications are all uses which are temporary or which involve regular replacement of the Guadua product, e.g. for maintenance purposes.

482 In Colombia the word '*esterilla*' is the common term for the bamboo mat. In other Latin American countries (e.g. in Ecuador) the term '*Guadua picada*' is used for this product. Other less common terms are '*Guadua abierta*', and '*Cana picada*'.

483 The '*cepa*', or bottom part of the Guadua culm, is usually sold in various lengths of between two and six metres, depending on the major application of this product in the construction sector. In the colder and higher areas of the study region (i.e. the department of Caldas), where the buildings are not constructed as high as in the lower and warmer areas (the departments of Quindio and Risaralda), the commercial '*cepas*' are shorter.

484 Information based on data compiled by HELD AND MANZANO (2004)

485 See *page xvi* for average exchange rates of the year 2002.

BERNAL[486] calculated an average annual volume of approx. 25 000m³ of harvested Guadua in the study region, with a value of more than US$ 3 million. Harvesting of the Guadua stands is controlled by the regional environmental authorities, the CARs. In the year 2002 a unified norm on the use and management of the Guadua stands in the study region was implemented; the '*Norma unificada en Guadua*[487]'. This norm replaced the different norms and regulations in place in the four study departments prior to 2002. The norm introduced a common volume calculation scheme[488] and a common fiscal code for utilisation fees. Four types of utilisation with a graded fee system have also been introduced. Utilisation for domestic purposes with harvested volumes less than 20 m³ annually are free of any fees. Nevertheless, such utilisation must be registered in the CAR offices (see *page xvi* for average exchange rates in the year 2002): Type 1 utilisation refers to harvested volumes up to 50 m³ annually for sale, and incurs a charge of Col. $ 2 500 per m³. Type 2 utilisation (volumes between 50 m³ and 200 m³ for commercial purposes annually) costs Col. $ 1 500/m³. Type 3 utilisation, which are commercially harvested volumes of more than 200 m³ annually, have to be planned by forest engineers and a management plan for the harvested Guadua stands has to be developed. The costs of the harvesting and management plan are covered by the stand owner. These plans must be approved by the CARs. The fee per harvested m³ is Col. $ 750[489].

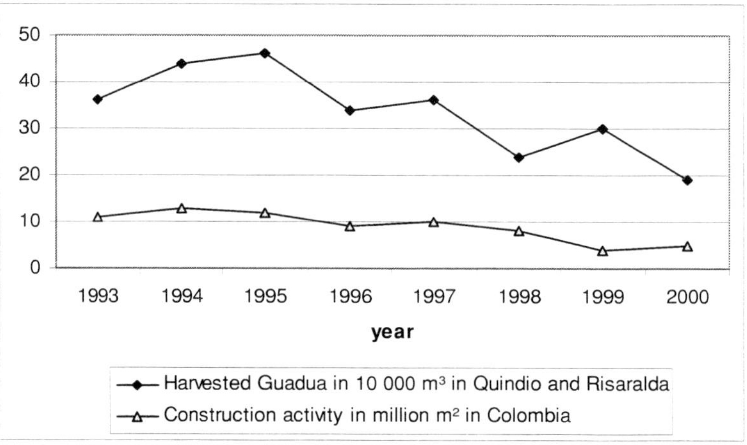

Graphic 1. Activity of the Colombian construction sector[490] and volumes of harvested Guadua[491] in the departments of Quindio and Risaralda between 1993 and 2000

[486] BERNAL (2002: 63f.)

[487] MSBC (2001 and 2003)

[488] This calculation scheme is based on experiments carried out by the CARs and declares that ten culms of a length of 6 m each are the equivalent of 1m³ of wood (MSBC, 2001: 10f.).

[489] MSBC, (2001: 10ff.)

[490] Internet database enquiry 2004: CAMARA DE CONSTRUCCIÓN DE COLOMBIA

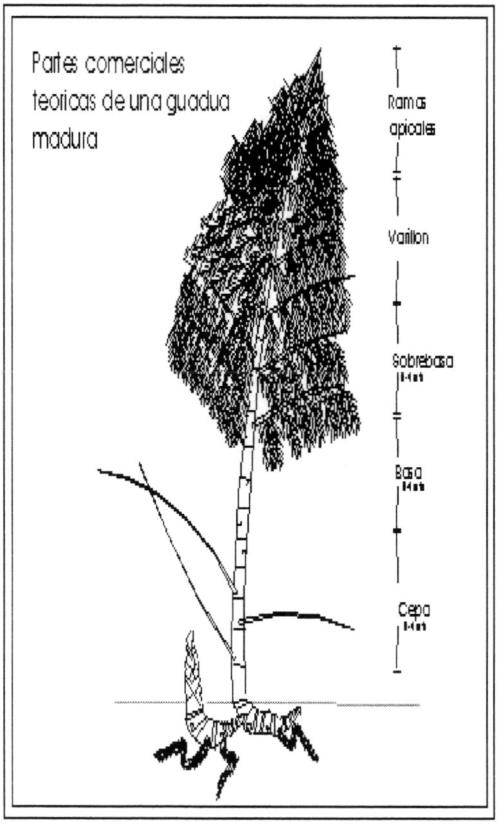

Figure 7. The Guadua plant and the commercial culm classification[492]

[491] Information based on data in BERNAL (2002); Note that the earthquake of January 1999 resulted in a counter-cyclic increase in demand for Guadua due to reconstruction activities in the study region.

[492] As in BUCHELLY (2001)

5. Research Methods

The critical discussion of the theoretical platform underlying this research, which was presented in the preceding chapter, resulted in a number of research hypotheses and questions cumulating in the following statement: "The manner in which actors interact determines how they innovate". Therefore, the research objects are the actors[493], their interactions[494] and the innovations[495] that come about subsequent to these interactions. The methodological fundament of this study was established by POPPER in 1934[496]. He pointed out that the falsification of existing theories is the major pursuit of scientific research, and that this process of falsification can be implemented through inductive empirical research, if the process of inductive deduction is open and free to interpretation and discussion[497].

To test the research hypotheses of this thesis, methodical approaches originating from empirical social and socio-economic research were selected. It was decided that the qualitative research approach best fits the demands of this kind of research, the foci of which are the actors' behaviour and interactions[498]. According to SCHNELL, HILL AND ESSER,[499] qualitative research is the compilation of techniques and methods[500] required to carry out scientifically sound and correct research on human behaviour and social phenomena, and this set of qualitative research methods finds application when theories on human action and interaction within a social context are subject to scientific analysis and evaluation. Therefore, qualitative social research aims at systematically comprehending and interpreting human behaviour and human perceptions, intentions, opinions, experiences and values[501]. In addition to the actors and their interactions, the actors' environment must be taken into consideration in the research methods, in order to better understand the actors investigated and to better interpret the research data and information, which in turn is necessary for the improved credibility of the results and conclusions derived from the research process[502].

---

[493] For the definition and classification of 'actors' see SCHARPF (2000: 86ff.) and *chapter* 3.3.

[494] For the definition and classification of 'interactions' see SCHARPF (2000: 90f. and 167ff.) and SCHIMANK (2002: 169ff.), and *chapter 3.3*.

[495] For the definition and classification of 'innovations' see OECD (1996: 10ff.) and *chapter* 3.2.

[496] In POPPER (1976)

[497] SCHNELL, HILL AND ESSER (1995: 58)

[498] MAYRING (1993: 35f.)

[499] SCHNELL, HILL AND ESSER (1995: 5f.)

[500] The difference between methods and techniques has been described by STATZ (2000: 101): A research method is the systematic, regular process of testing and verifying theories, whereas the research technique is the very specific mode of operation within a methodical process.

[501] ATTESLANDER (1995: 11ff.)

[502] MAYRING (1993: 9)

The selected set of research methods has to meet quality requirements in order to ensure the expected level of reliability and the validity of the research results. The following quality requirements were compiled by FLICK (1996)[503] and MAYRING (1993)[504]:

Documentation of the research process/Procedural reliability and validity
Detailed documentation of the applied research methods and techniques, as well as the detailed description of the research and its interpretation context, in order to make the research process repeatable and the research results interpretable for other researchers.

Argumentative testing and warranting of the interpretation/Analytical induction
Interpretation of the research results from various alternating points of view and under different interpretation contexts. The rejection of as many interpretation cases as possible underlines the validity of the results.

Compliance with the rules of the research process
The research has to follow systematic steps and phases of data collection and analysis. In these single steps/phases the quality requirements can be realised more readily and efficiently. Errors, misinterpretations and other problems can be detected earlier and more easily, and the whole research process becomes more transparent.

Closeness to the research subject
In order to produce first-rate qualitative research results, the researcher has to conduct the research as close to the research objects as possible. As qualitative research basically investigates social environments, inter-human relations and the actors' behaviour, the researcher has to meet the subjects at an equal and emancipated level.

Communicative validation
The research results should be discussed with the research objects themselves, in order to verify the interpretation context and to complete the argumentative testing of the research data and the research results. This can happen in workshops, in group discussions or in an individual dialogue with selected subjects.

Triangulation
Triangulation refers to all levels of the research process. It involves the diversification of theoretical frameworks, data sources, research objects, researchers, interpretation contexts, analysis tools, etc. The aim of triangulation is to avoid subjectivity in data collection, data analysis and in the interpretation of the results through the application of various possibilities, which may then provide the most reasonable solutions and answers.

---

503 FLICK (1996: 239ff.)

504 MAYRING (1993: 106ff.)

The qualitative research approach covers mainly the actors' behaviour and interactions. The perceptions and attitudes of the actors are surveyed and analysed. A quantitative research approach, by contrast, usually measures or counts data and information on, for example, socio-demographics, economics or politics. The differentiation between qualitative and quantitative research may be located at various levels. The difference may be identified in the whole research approach, in the research methods applied or in the data gathered. *Table 19* highlights the most important levels of differentiation between the attributes 'qualitative' and 'quantitative' in social science research. In this study, the overall qualitative research approach also applies research methods to the collection of quantitative data. Quantitative methods and data operate at a higher level of abstraction than qualitative methods and data. The application of such quantitative methods in qualitative research approaches reduces the complex sociological data and improves the interpretation context[505].

Table 19. Characteristics of 'qualitative' and 'quantitative' in social science research[506]

	'qualitative'	**'quantitative'**
Approach	basic constructivist belief inductive approach	basic positivist belief deductive approach
Methods	Open and unstructured methods case studies, different interview and observation techniques	Standardised and structured methods, e.g. experiments, surveys, secondary data analysis, content analysis
Data	words, drawings, numbers, etc.	abstract data: usually numbers

For many years the contrasts between qualitative and quantitative research methods have been emphasised in the methodological discussion. However, in recent years it has become obvious that both methods account for significant structural deficits if used isolated from one another. Authors and researchers are increasingly emphasising the positive synergetic effects that a combination of methods from the two backgrounds provides, and that this can help to overcome the deficits of each individual method[507].

SILVERMAN[508] wrote that the combination of qualitative and quantitative methods enables the researcher to work on a large amount of data without getting inundated, and MAYRING[509] pointed out that the combination of methods from both methodological backgrounds makes possible improved qualitative content analysis, improved validation and improved generalisation of the research results and a better research design. He pushed forward the application of quantitative methods in the analysis of qualitative data, in particular (see

---

505 KLEINING AND WITT (2001: 76)

506 As compiled from SILVERMAN (2001: 25ff.)

507 FLICK (1996: 10f.); MAYRING (2001: 1ff.)

508 SILVERMAN (2001: 35ff.)

509 MAYRING (1993: 24f. and 2001: 8ff.)

chapter 5.3). KELLE[510] sees in the combination of qualitative and quantitative methods the consequent compliance with the quality requirement of 'triangulation'. He argues that the application of both methodological approaches to the investigation of the same scientific context should be an obligatory step in every qualitative research design. Whereas MAYRING[511] emphasises the importance of the combined methods in a research process, where one method follows the other, KELLE AND ERZBERGER (2000) see the strength of the combination in the parallel application of both methods.

In MAYRING'S argumentation, the quantification of qualitative data is a tool to improve interpretation, validation and generalisation of the same information. KELLE[512], on the other hand, recognises the possibility that the application of quantitative and qualitative methods to the same research problem may result in different interpretations and analysis results, which of course also contribute to better validation and greater credibility of the research results. These different interpretation contexts may be:

- Qualitative and quantitative results may converge: in this case the results lead to the same conclusions.
- Qualitative and quantitative results may relate to different objects or phenomena, but may be complementary to one another and can, therefore, be used to supplement each other.
- Qualitative and quantitative results may be totally divergent or contradictory.

The research design for this study, as presented in *figure 8*, applies a combination of both qualitative and quantitative research methods within the framework of an overall qualitative approach in form of a case study. The different phases and steps of this design will be explained in the following chapters.

---

510 KELLE (2001: 6ff.)

511 MAYRING (1993: 24f.)

512 KELLE (2001: 15ff.)

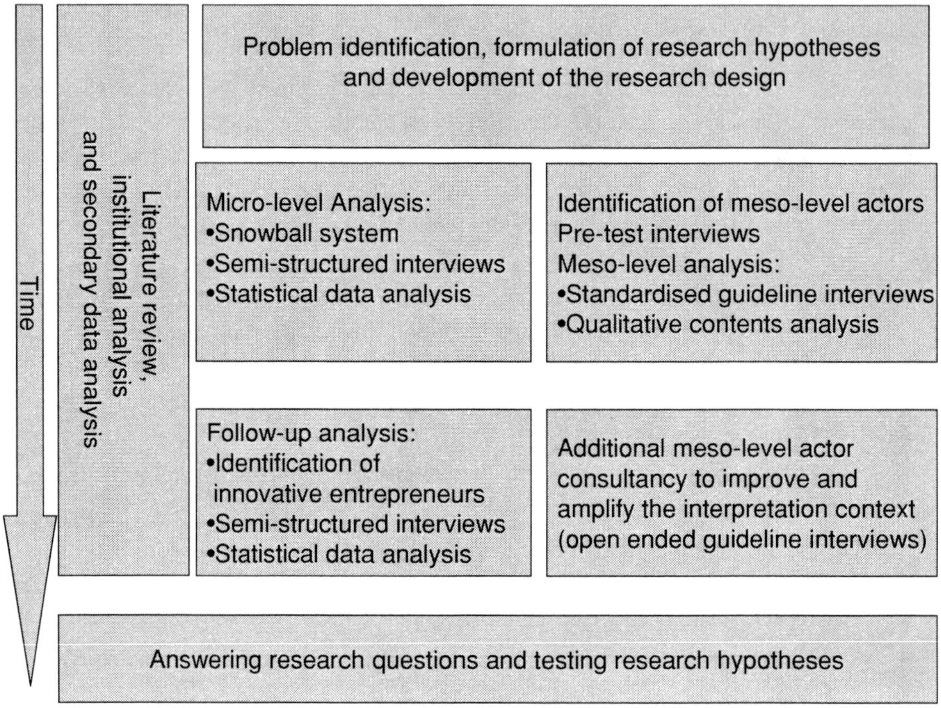

Figure 8. Research design as applied to this study

5.1 Secondary Data and Information

In preparing for the primary data collection, the analysis of the data and the interpretation of the final results, a number of secondary data sources, literature and other sources of information were revised and evaluated in terms of their quality and possible value in relation to this research project. In the following section the most important literature, secondary data sources and information for the accomplishment of this study are presented.

One major source of information during all of the research phases were the regional environmental authorities[513] and the ministry for the environment[514], the ministry for industry and export[515] and the ministry for agriculture[516]. They provided official information on the legal status of the bamboo resource base and the related actors. They provided material in

[513] Span.: *Corporaciones Autonomas Regionales (CARs)*

[514] Span.: *Ministerio del Medio Ambiente (MMA)*

[515] Span.: *Ministerio de Comercio Exterior e Industria*

[516] Span.: *Ministerio de Agricultura y Desarrollo Rural*

relation to the laws, official regulations and other general information. Further, they keep statistics on: the harvested and traded volumes of bamboo raw material, structural data for the manufacturing sectors and national export data. For detailed information on the management and use of the Colombian forests and bamboo resources, the national forest project MSBC[517] was consulted, which provided statistical information and maps. Detailed statistical information on departamental and municipal structural data and GNPs, as well as socio-demographic information, could be requested by means of an internet database enquiry from the national Colombian statistical department (DANE[518]). Another important source of economic and social information on the coffee growing region was the Colombian coffee growers' association (FEDECAFE[519]).

One important limitation of the information and sources mentioned above was the lack of data concerning the informal sector and micro-enterprises. For the bamboo sector, such information was provided by a study of the Colombian bamboo production-to-consumption-system by HELD AND MANZANO (2004). In addition to this study, a number of scientific publications and research documents on all aspects of bamboo management, use and commercialisation were revised. This information could be found easily thanks to a scientific register of bamboo related documents from 1950 to 2000 prepared by ARBELAÉZ (2001). In order to place the Colombian bamboo sector in an international context, valuable information was provided in the form of case studies conducted by INBAR in various countries throughout the world, the INBAR database and the INBAR Latin American office in Ecuador. However, the focus of this organisation remains the Asian continent, with its tremendous bamboo resources. Information on Latin American countries is scarce.

To get in touch with experts and to follow the process of institutional and technological developments in the Colombian bamboo sector, memories were probed and conference publications of a number of events held in Colombia and Latin America were revised. These sources provided useful information on experts and the processes, which have taken place so far. Other important sources of information were the national Colombian centre for investigations in bamboo[520] and the president of the Colombian bamboo society[521], Ximena Londoño, who also acted as gate keeper during the preparation and performance of the meso-level actor interviews (see *chapter 5.2.2*).

[517] Span.: *Manejo Sostenible de Bosques en Colombia*

[518] Span.: *Departamento Administrativo Nacional de Estadistica*

[519] Span.: *Federación de Cafeteros de Colombia*

[520] Span.: *Centro Nacional para Investigaciones en el Bambú Guadua*

[521] Span.: *Sociedad Colombiana del Bambú (SCB)*

5.2 Primary Data Collection

The definition of the scope of the research and the type and number of research objects to be included in it, is one of the most crucial phases of the research process. The identification of research objects usually involves 1) the definition of the criteria, which must be met by the research objects in the overall research ('target population'), and 2) the determination of a research sample, when the total of the target population is unknown or too large to be investigated in its entirety.[522]

The theoretical framework supporting this research project, which defines the Guadua sector as an economic survival cluster, provides the criteria for the identification and definition of actors to be included in the target population (see also *chapter 2.2* on clusters). Basically the research objects (=actors[523]) can be classified into meso-level actors and micro-level actors.

This rough classification of the total research objects has to be refined later on, in order to ensure the collection of data for as many actor groups in as many interpretation contexts as necessary. Here, the actors at the micro-level were classified according to their economic activity as entrepreneurs in the Guadua sector. In the meso-level the actors were classified according to their institutional background (*table 20*). This chapter is structured into the methodological approach of the identification and data collection at the micro-level and the methods applied to identify and investigate the meso-level actors.

5.2.1 Meso-level Analysis

The meso-level actors are actors, whose activities intentionally influence sectors of societal interest, such as culture, education, public security, health or economic sectors[524]. In the context of this study, the Guadua sector of the Colombian coffee growing region, as described in *chapter 4*, represents such a sector of societal interest. The meso-level actors affect this sector through, for example, export promotion policies, environmental legislation, extension activities, providing funds and credit schemes, facilitating technological transfer, etc. Usually, they are government and non-government actors, but also actors originating from public-private[525] co-operations and private actor organisations. The meso-level is the level of specific sector policy-making and is located between the macro-level of large scale national policy-making and the micro-level of enterprises[526]. For the purpose of this research, all of the Guadua-sector related meso-level actors were desired for inclusion in the research process.

---

[522] FLICK (1996: 78ff.) and SCHNELL, HILL AND ESSER (1995: 251ff.)

[523] In this study, actors are defined according to SCHARPF (2000: 86ff.; see also *chapter 3.3*)

[524] As adopted and modified from MEYER-STAMER (2001)

[525] In the course of the present study, public-private actors are those, which receive funds and financing from both sides and which provide partly public and partly private services to their target groups/clients. Such public-private actors/organisations are for example the chambers of commerce, collaborative R&D institutes or co-funded initiatives.

[526] MEYER-STAMER (2001: 2 and 7ff.)

The actors could be identified by means of the participation lists of the most important Guadua related activities and events held in the study region in the year 2001. This methodological approach belongs to the set of subjective selection of research objects, which are discussed later on in this chapter.

Besides the actors, who actively and intentionally influence the sectors of societal interest mentioned above, there are actors, who influence these sectors without having any such intention, or without even realising that they do so. These actors, per definition, do not belong to the meso-level group of actors in this study. However, it is necessary to systematically analyse these actors in order to describe and weigh their impact upon the sector of interest. In this case such actors were tracked and identified when named by one of the interviewed actors from the micro and meso-level. Contact with these actors was attempted and interviews with them were sought.

Most of these actors could be identified and were interviewed during the second phase of the research process. Such interviews were conducted in an unstructured and open-ended manner. The purpose of these interviews was to identify the impact the actors have on the Guadua sector. In other cases, the interview topic was the 'non-impact' of the interviewed actors on the sector, although it was assumed that they do have one. These interviews were analysed individually and the results were used to amplify the context of the interpretation of the meso- and micro-level analysis. Furthermore, the analysis of such actors was indirectly conducted during the preparatory and accompanying institutional analysis, as such actors frequently effect the sector through institutional arrangements, for example, through official laws and regulations, informal codes of trade and payment, or the influence they have on trust and mistrust in the society.

In this study actors influencing the Guadua sector, but not included in the systematic meso-level analysis through semi-standardised guideline interviews (see *chapter 5.2.2*), were identified from the timber producing and processing sectors, from the pulp and paper industry, from the *National Innovation System* and the public and private banks. Interviews could be conducted with representatives of these groups. Another group of such actors was identified belonging to armed conflict parties, such as the various guerrilla organisations (*FARC, ELN, etc.*[527]), the paramilitary organisations (*AUC*[528]), and those involved in the production of illicit drugs, but also the legal governmental units of the police and the army. All of them affect greatly the social and economic life in the study region. No interviews have been conducted with representatives of these organisations.

The selection process, as it was applied for the identification of the meso-level actors, is described in the literature as subjective selection. SCHNELL, HILL AND ESSER wrote on the subjective selection of research objects that this method of selection is useful for expert

---

[527] Acronym.: *FARC* = Fuercas armadas revolucionarias de Colombia; *ELN* = Ejercito de la liberación nacional

[528] Acronym: *AUC* = Autodefensas Unidas de Colombia

interviews and for the selection of both 'typical' and 'extreme' cases[529]. At the meso-level the research objects were organisations rather than the interviewed actors themselves. These organisations were selected according to their engagement in the Guadua sector. However, within the respective organisation an interview partner knowledgeable 1) in the specific field of research and 2) in the organisation itself had to be found. The interview partner had to fulfil the role of an expert, as well as that of a representative of the organisation[530]. DEEKE[531] defines an expert as a person with special, almost exclusive knowledge in a certain field of interest, and goes about with this knowledge responsibly and reasonably.

According to MAYRING[532] qualitative research methods are the most appropriate tools to analyse and evaluate action and interaction oriented research aspects. As the appropriate instrument for the collection of research relevant data and information, the personal interview method was chosen, applying the face-to-face questionnaire interview technique. To conduct such a research interview a set of scientific criteria has to be fulfilled to ensure the scientific soundness of this method[533]. The exercise of such interviews has to be...:

- systematically prepared and conducted,
- objective oriented and
- theoretically embedded and controlled.

In addition to the compliance with these quality criteria, the researcher has to consider the character of the interview situation itself as an influencing and determining factor for the results of the research. Qualitative interviews work according to the stimulus-reaction model, which describes the interview situation as a reaction system. The most important aspects of this model are: the notion that there is no correct answer; reactions depend on the interview context and environment; answers are influenced by the perceptions and expectations of interviewer and interviewee; the interpretation of the interview is part of the answer; and the interview can only be one part of a set of qualitative methods to investigate in one research context (ATTESLANDER, 1995)[534]. In order to conduct this research study two types of face-to-face questionnaire interview techniques were selected:

- The standardised semi-structured interview for the micro-level actors (*chapter 5.2.5*)
- The semi-standardised guideline interview for the meso-level actors (*chapter 5.2.2*)

---

[529] SCHNELL, HILL AND ESSER (1995: 281ff.): 'Typical' cases are cases characteristic of the target population and represent the whole set of the target population's characterisation criteria, whereas 'extreme' cases consist of single, extremely developed characteristics of a target population.

[530] See MIEG (1994a) on the role of experts and DEEKE (1995: 8ff.) on the definition of experts and the implementation of expert interviews

[531] DEEKE (1995: 8f.)

[532] MAYRING (1993: 35ff.)

[533] ATTESLANDER (1995: 132ff.)

[534] ATTESLANDER (1995: 136ff.)

5.2.2 Meso-level Guideline Interviews

During the course of the research 43 meso-level actors were interviewed during the face-to-face semi-standardised guideline interviews. The meso-level actors interviewed were classified into 6 groups according to their organisational background (see *table* 20).

Table 20. Meso-level Actor Groups

Actor Group	Number	Percent
Government organisations	13	30
NGOs	4	9
Construction sector	5	12
Agricultural producer organisations	3	7
Research entities	7	16
Independent experts	6	14
Combined private-public actors	5	12
Total	**43**	**100.0**

The semi-standardised guideline interview is characterised by pre-formulated open-ended interview questions on specific research topics. With this type of interview, the interviewee is encouraged to explain his point of view on the research topic/problem and contribute to the falsification/verification of the theory guided research hypotheses. The questions in this type of interview are open-ended questions on the research problems, structured into thematic fields, theory guided questions and confrontational questions[535].

Conducting a guideline interview with a meso-level actor requires compliance with requirements to ensure the scientific soundness of the results of such an interview. MIEG (1994a and 1994b) recommends that an interviewer…:

- makes use of 'gate keepers[536]', who may recommend experts/interview partners and help to interpret their interviews,
- is scientifically and professionally prepared,
- transforms the relevant research hypotheses and questions into a guideline questionnaire,
- does not force the interview partner to comment on fields other than that of his expertise and
- uses the most complete form of interview documentation, namely tape recordings.

[535] FLICK (1996: 100f.): The confrontational questions provide alternative solutions to the research or theoretical problem, which the interviewee has to anticipate and position himself against.

[536] MIEG (1994b: 13): 'Gate keepers' are people extremely knowledgeable amongst a certain circle of experts or in a thematic field. They may establish contacts and may comment and interpret the interview behaviour of the experts. The 'gate keepers' themselves are not experts!

To identify the meso-level actors (organisations) relevant to the accomplishment of this study and to contact the appropriate interview partners within these organisations, the president of the Colombian Bamboo Society, Ximena Londoño, was incorporated into the research scheme in a gate keeper function. She established the necessary contact with the organisations and on the basis of her 20 plus years of experience in the Colombian, Latin American and worldwide bamboo sectors, she could recommend interview partners within these organisations knowledgeable in relation to the topics relevant to the research, people normally unknown to the foreign researcher. Furthermore, she acted as a consultant in biological and taxonomical questions arising during the work in the Guadua sector.

Other measures recommended to ensure the soundness and validity of the expert interviews were achieved by making audio recordings of the interviews and the immediate transcription of the tapes. During the interviews, the field of expertise of each interviewee was respected. However, during the interview partner selection process, candidates who would be able to comment on all fields of the research related topics were sought.

Subsequent to the research problems and hypotheses presented in *chapter 3.4*, the following guideline topics for the meso-level actor interviews were designed. These topics were structured into four thematic fields. The interview questions related to these fields were ordered from introductory to detailed questions.

Regional development

- Basic understanding of regional development
- The situation of the study region in the context of regional development
- The situation of the bamboo resource and its use in the study region

Actor resources[537]

- The actor's major resources
- Resource oriented strategies
- Strengths and deficits in relation to resource endowment

Actor strategies and interactions[538]

- Strategic self-positioning in the bamboo sector and the related policy fields
- Participation in the decision-making process
- Co-operation and communication with other actors

[537] The actor's resources were defined in *chapter 3.3* according to SCHARPF (2000: 86f.)

[538] Interactions in the context of the research were defined in *chapter 3.3* according to SCHARPF (2000: 90f. and 197ff.)

Development and promotion of the Guadua sector

- Actor's aims for the Guadua sector
- Sector promotion activities
- Problems and obstacles in the sector

5.2.3 Meso-level Questionnaire Interviews

The qualitative data collected through the guideline interview was accomplished by means of a set of structured interview questions. These questions covered similar topics to the guideline questions in the first part of the questionnaire. The aim of this activity was to gather countable data relating to actors strategies, activities and interactions for further interpretation of the qualitative data and information from the first part of the interviews. The elaboration of this part of the interview questionnaire was based on research conducted by MANSKE, RUTH AND DEITMER in 1999[539], which focused on the evaluation of meso-level actor activities with respect to sector and innovative promotion of an industrial region in Germany. This approach was selected because, within the framework of innovation and growth theories, MANSKE, RUTH AND DEITMER put the actors and their interactions at the forefront of the research interest[540], which corresponds entirely with the objectives and the background of this study.

The structured interview questions contained in the actor questionnaire aimed at the self-evaluation of the actors' activities and interactions. The questions were structured as a ranked inventory of the actors' activities on a scale of 1 to 5 ('1'=lowest intensity of activity, '5'=highest intensity of activity, '0'=no activity in the specific field of action). The interviewees were asked to evaluate their activities in the various fields of activities for the years 1997 and 2002, and to give a prediction for the year 2007 (see *table 21*).

List of bamboo sector related activities covered by the questionnaire:

- Extension in business affairs
- Technical extension
- Professional education
- Basic biological research
- Silvicultural research
- Product research and development
- Process research and development
- Market research

---

539 MANSKE, RUTH AND DEITMER (2000)

540 MANSKE, RUTH AND DEITMER (2000: 5ff.)

- Participation in Guadua related associations
- Participation in institution building
- Funding of bamboo projects
- PR and participation in exhibitions
- Other activities mentioned by the interviewee

Their activities were evaluated by the interviewees on a scale of 1 to 3 ('1'=results/outcomes are not satisfying and do not justify the resources expended, '2'=results/outcomes are satisfying and justify the resources expended, '3'=the results are very satisfying and expenditure of resources is absolutely justified)[541].

Table 21. Questionnaire excerpt: Actor's activities and self-evaluation

	Importance of activity (please mark only once for each year and activity)					**Evaluation of results (please mark only once for each year and activity)**		
Activity	1 (very low)	2 (low)	3 (medium)	4 (high)	5 (very high)	1 (un-satisfying)	2 (satis-fying)	3 (very satisfying)
Silvic. research 1997								
Silvic. research 2002								
Silvic. research 2007								

The last part of the questionnaire surveyed and evaluated the actor's level of external interaction, again at a scale of 1 to 3[542]. This data set referred to co-operation with other actors, the areas in which such co-operation took place and evaluation of the outcomes according to the resources expended (see *table*s *22* and *23*).

[541] As the interviewees frequently answered in between the given ranks, intermediate ranks were established between '1' and '2', and '2' and '3' for the purposes of the data analysis, resulting in a scale of 1 to 5. During the analysis of the data the order of ranks was reversed so that (just like the style of German school grades) '1' is the best valuation and '5' the worst.

[542] For the significance of this scale see preceding reference.

Table 22. Questionnaire excerpt/Example of data collection: Actor co-operation

Partner in co-operation	Year of initation	Area of co-operation	Results of co-operation	Perspectives for co-operation
Partner A	2000	Market research	Report on European flooring market	Co-operation completed
Partner B	1999	Flooring research	Prototype of a lath saw	Funding ended, but continuation in planning
Partner C	2001	Export promotion	None yet	Until 2003

Table 23. Questionnaire excerpt: Actor evaluation of co-operation

Type of co-operation	Evaluation of co-operation (1=not satisfying; 2=satisfying; 3=very satisfying; Please mark only once for each partner and co-operation type)								
	Partner A			Partner B			Partner C		
Financial co-operation	1	2	3	1	2	3	1	2	3
Technical co-operation	1	2	3	1	2	3	1	2	3
Staff co-operation	1	2	3	1	2	3	1	2	3
Institutional co-operation	1	2	3	1	2	3	1	2	3

Table 22 establishes the co-operation links to other actors for the purposes of survey and evaluation. The four areas of co-operation referred to are financial, technical, staff and institutional co-operation. Financial co-operation refers to the financing of projects and any other kind of monetary contribution to other actors, their research and activities. The area of technical co-operation was characterised by the exchange of technical equipment and knowledge. Staff co-operation was any kind of temporary or permanent exchange of employees or experts between the respective co-operation organisations/actors, as well as the creation of combined project and research groups. Finally, institutional co-operations were described as the co-operations aiming at the creation of any kind of institutions, such as legislation, construction standards, utilisation guidelines, etc. Usually, this kind of co-operation was achieved during regular meetings of the co-operation partners at round table discussions or project groups.

5.2.4 Micro-level Analysis

The micro-level actors are entrepreneurs actively engaged in the Guadua production-to-consumption system (PCS) of the study region. The PCS is defined as: "*the entire set of actors, materials, activities and institutions involved in growing and harvesting a particular raw material, transforming the raw material into higher-value products and marketing the final product*[543]". The definition of the classification criteria for the entrepreneurs was elaborated according to HERRERA'S guide to the elaboration of an actor typology (1999). They are producers of raw material, processors, intermediaries, wholesalers or retailers. Experts were also included in the research of the micro-level, for example knowledgeable Guadua architects and engineers. As the total size of the target population at the micro-level was unknown, the following selection methods were used for the purposes of this research: 1) The snowball system to identify producers, processors and dealers[544]; and 2) the subjective selection of 'typical' actors[545]

The snowball system was described by SCHNELL, HILL AND ESSER as a way to analyse networks (of families, conflict parties, enterprises, etc.). It makes it possible to approach a large number of research objects step by step. Of course, the system is not a random sampling method, but a subjective conscious selection. In the case of this study, the starting point of the research was an exhibition of handicraft producers in the centre of the major city within the study region.

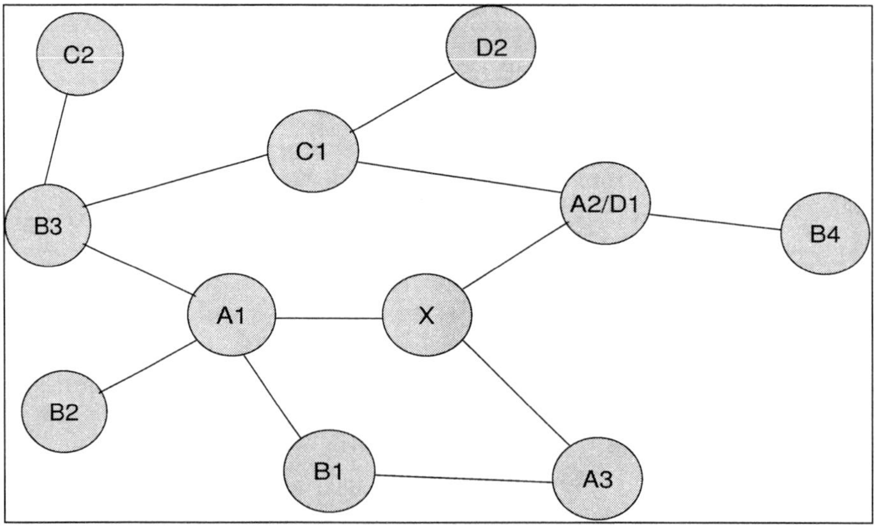

Figure 9. Snowball system for the selection of research objects in target populations of unknown size

[543] BELCHER (1998: 59)

[544] SCHNELL, HILL AND ESSER (1995: 281f.)

[545] SCHNELL, HILL AND ESSER (1995: 281) and as described in *chapter 2.2.1.*

The above graph illustrates the selection approach of the snowball system. The basic idea is to make use of the research objects' knowledge and contacts, for example, to ask a limited number of members of the target population (in the figure above this is member 'x') for the names and addresses of other members of the target population. In this study the questionnaire designed for the handicraft and furniture producers in the bamboo sector included one question asking for information concerning competing and co-operating enterprises in the sector. When conducting interviews with the producers named by the interviewee 'x' (marked as 'A1, A2, etc.' in the figure) they in turn were asked for other members of the target population, and so on.

The system results in concentric circles of new research objects around interviewee 'x' (A-circle, B-circle, C-circle, etc.). The number of newly mentioned enterprises per interviewee declines with every circle-level, because contacts are repeated (see *figure 9*: interviewee x names A2, as well as C1, therefore enterprise A2 is also D1) in each concentric circle level. This of course depends on the total size of the target population and the intensity of interactions the members of the target population maintain with each other (GABLER, 1992).

5.2.5 Micro-level Questionnaire Interviews

Standardised semi-structured interviews were conducted with the members of the micro-level identified as described in this chapter. The standardised semi-structured interview consists of sets of pre-formulated questions[546]. In this study, sets of pre-formulated open questions and structured closed questions[547] were combined. This type of interview was applied for micro-level entrepreneurs and meso-level actors. Therefore, only the second part of the meso-level actor interviews was structured (see *chapter 5.2.3*). The first part of the meso-level actor interview was a guideline interview (see *chapter 5.2.2*).

The research at the micro-level was conducted in two phases: one baseline-study in the year 2002, which covered a broad range of micro-economic and institutional aspects (see *table 24*). In this research phase four different questionnaires were designed for four entrepreneur groups (see *table 26*) basically covering the same topics, but also branch specific details. As the most important entrepreneur groups in this study were the handicraft and furniture producers, the examples and explanations in this methodology chapter refer to their questionnaire. A second follow-up interview campaign was conducted 18 months later (2004), which covered special topics concerning the innovation processes in the same handicraft and furniture enterprises investigated in 2002 (see *table 25*).

Quantitative and qualitative data relating to innovations was gathered by means of the micro-level interviews. This activity focused on the micro-economic standard data relevant to the measurement of the innovation of the enterprises. The quantitative data collection was carried

546 ATTESLANDER (1995: 162f.) and SCHNELL, HILL AND ESSER (1995: 299f.)

547 ATTESLANDER (1995: 180ff.) and SCHNELL, HILL AND ESSER (1995: 303ff.)

out mainly in the form of structured questions in the questionnaire, which were combined with semi-structured and open-ended questions referring to the qualitative data collection (see questionnaire no. 2 in the *annex* 2).

The theory of innovation, its importance to economic growth and the problems researchers have encountered so far in explaining the role of interactions in the innovation and diffusion process are described and explained in *chapter 3.2*. To draw conclusions on the role of interactions in the innovation process, quantifiable parameters are required, which give evidence of the innovation activities of the research objects. The measurement of innovation and innovative activities can be achieved by counting the number of patents handed in per firm, productivity per worker, product and technology replacement ratio, expenditures on R&D, etc. [548]

This study assesses the contribution of newly adopted products to the enterprises' sales volume, the ratio of technology acquisition and the changes in the organisational structure of the enterprises[549] (see *table 25*). Furthermore, this research aims to investigate the sources of knowledge behind the innovations adopted[550] and the costs of the innovation process[551]. The decision on the parameters, which had to be assessed, was supported by the OECD Manual (1996) on the measurement of scientific and technological activities.

Table 24. Questionnaire structure: micro-level entrepreneurs PCS-study 2002

Question topics	Type of questions
Basic enterprise data	Structured
Product/raw material acquisition	Structured closed and open
Transformation/manufacturing process	Structured closed and open
Final products	Structured closed
Marketing strategies	Structured closed and open
Machinery/technology	Structured closed
Employment structure	Structured closed
Sales volumes	Structured closed and open
Client structure	Structured closed
Foreign capital acquirement	Structured open
Horizontal interactions	Structured open
Vertical interactions	Structured open
Snowball information	Structured open
Institutional arrangements, setting and support	Structured open
Plans and projections	Structured open

[548] OECD (1996: 49ff.)

[549] OECD (1996: 53ff.)

[550] OECD (1996: 51f.)

[551] OECD (1996: 60ff.)

Table 25. Questionnaire structure: micro-level entrepreneurs innovation study 2004

Question topics	Type of questions
Basic enterprise data	Structured
Sales volume	Structured open
Employment structure	Structured open
New products	Structured open
New machinery	Structured closed and open
New clients/markets	Structured open
New organisation	Structured open and closed
New capital sources	Structured open and closed
Sources of knowledge for innovative products	Structured closed and open
Product innovation activities	Structured closed and open
Product innovation costs	Structured closed
Diffusion paths	Structured open
Institutional setting	Structured open

On the micro-level, semi-structured interviews were conducted with 113 entrepreneurs, as shown in *table 26.*

Table 26. Micro-level entrepreneurs interviewed in 2002

Entrepreneur group	Number
Special providers of raw materials and products	5
Civil engineers/architects	18
Intermediaries/distributors/retailers	30
Handicraft producers	48
Furniture producers	12
Total	**113**

For the implementation of the interviews with the micro-level actors, a Colombian student was integrated into the research. At the beginning of this research phase, the student accompanied the researcher during the micro-level interviews. Later, the student conducted the interviews while the researcher observed the situation without interrupting or intervening. The course of these interviews was discussed the same day. The student finally conducted more than fifty of the interviews alone. The high number of interviews (twenty) conducted jointly was necessary to ensure the scientific soundness of the interviews and their documentation. This instruction phase was crucial in order to guarantee the interview results were of the same quality over time, and to homogenise the results of interviews by various interviewers, in this case the researcher and the assistant[552].

[552] For further information on the instruction of assistant interviewers see Atteslander (1995: 198ff.)

5.3 Data Analysis

The analysis of the data and information gathered through the aforementioned methods was differentiated into the analysis of the qualitative data originating from the semi-standardised guideline interviews and partly from the semi-structured interviews with the micro-level entrepreneurs, and the analysis of the quantitative data resulting from the structured interviews of the meso-level actors and the entrepreneur. The analysis of the qualitative data was carried out according to MAYRING'S qualitative content analysis[553], whereas the analysis of the quantitative data was by means of uni- and bivariate statistics using MICROSOFT EXCEL and SPSS 10.0 software[554].

5.3.1 Qualitative Content Analysis

The content analysis is a method used to analyse the contents of communication[555], which originally involved the counting of motives or statements and other quantitative measurement of the text contents in a huge number of comparable text bodies, such as newspaper articles or standardised interviews[556]. This quantitative approach to the analysis of text content was amplified through qualitative analysis methods, which focused on the context between text passages, the text meaning, the concentration on individual texts and the 'unwritten' contents[557]. The qualitative content analysis is the most appropriate and most common form of analysis of qualitative data today[558]. For the purpose of this study the qualitative content analysis was chosen as this method was expected to fit the type of data gathered and complies with the quality requirements demanded of the methodological research approach.

The analysis of guideline interviews, as they were conducted for this study, through content analysis requires the interview transcripts. The analysis of the interview contents is the analysis of the dyadic interaction between the interviewer and the interviewee[559]. This interaction has to be comprehensible to and interpretable by the interview analyser. The guideline interviews were recorded with a dictaphone and transcribed later on the same day, as recommended by MIEG[560] and FLICK[561].

---

[553] MAYRING (1993: 85ff.; 2000 and 2001)

[554] The application of the SPSS 10.0 software was guided by the handbook by BÜHL AND ZÖFEL (2000)

[555] In this case communication is defined as any kind of information manifested in a text (MERTEN, 1995: 16)

[556] MAYRING (1993: 85) and MERTEN (1995: 14ff.)

[557] MAYRING (1993: 85f.)

[558] ATTESLANDER (1995: 246)

[559] MERTEN (1995: 275)

[560] MIEG (1994b)

[561] FLICK (1996: 186ff.)

The transcription was conducted thematically, in this case. The tapes were listened to at least twice and the content was summarised under the respective guideline questions[562]. This step simultaneously involved the categorisation of the information gathered. A list of the established categories is attached in *annex 4*.

According to MAYRING, the qualitative content analysis (QCA) is characterised by its controlled methodical approach and its step-by-step analysis process, which allows the segmentation of the text body[563]. For the purpose of the interview analysis, the contents of the text bodies were analysed according to the process of inductive categorisation[564], which combines the reduction of text volume with the amplification of text contents.

Categorisation is the first step of quantification of qualitative data in the data analysis process. The categories in the inductive categorisation process result from the text contents only and reflect the answers in the text (interview) to the research question. In contrast, during the deductive categorisation process the categories are pre-formulated and the text is attributed to these categories[565]. However, in both cases qualitative data is categorised and quantified for further quantitative analysis. In the last step the results are qualitatively interpreted (*figure 10*).

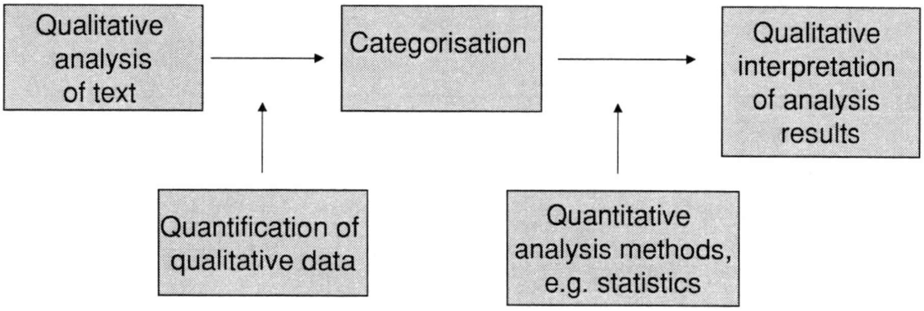

Figure 10. Basic process of the qualitative content analysis

The categorisation procedure of the guideline interviews resulted in the creation of categories and subcategories of actor statements in response to the guideline questions. FLICK describes the categorisation as a process of thematic codification[566]. This process is an analysis method, which identifies differences and commonalities between the actors in the target population in relation to a specific research item[567]. MAYRING integrates the thematic codification into the

---

562 FLICK (1995: 192f.) and SILVERMAN (2001: 163ff.)

563 MAYRING (1993: 86)

564 MAYRING 2000: 9ff.)

565 MAYRING (2001: 14ff.)

566 MERTEN (1995: 320ff.) recommends the development and use of a codification guide to categorise contents.

567 FLICK (1996: 210f.)

qualitative contents analysis procedure[568]. In order to analyse the categories of the guideline interviews, the data is numerically codified and entered into a data matrix, which makes possible the quantitative analysis of the data (see *chapter 5.3.2*).

5.3.2 Data Matrix and bivariate Statistics

The categories established through the qualitative contents analysis were codified and entered into SPSS 10.0. The statements were transformed into binomial variables, classified '0' and '1', with '0' signifying that the statement was not mentioned by the interviewee and '1' that the statement was mentioned by him/her[569]. *Table 27* shows an example of a data matrix resulting from the input of the categorised and codified interview data. This data was analysed using descriptive uni- and bivariate statistical approaches for binomial sets of data in SPSS 10.0.

Table 27. Example of a data set in SPSS 10.0

ID of inter-viewee	Guadua initiation	Actor group	Obstacles to co-operation	Obstacles to co-operation	Obstacles to co-operation
			Communication gap	Perceptions of competition	Lack of personal resources
CM01	1987	2	1	0	0
CP13	1999	6	0	1	1

In the data matrix for the quantitative data from the meso-level and the micro-level interviews, the figures and units were entered as they had been noted in the questionnaire forms during the interviews.

In addition to the basic univariate statistical analysis of the quantitative data (e.g. mean values, standard deviations, etc.[570]), the codified qualitative data was analysed by means of bivariate statistics[571]. The most important analysis method was the cross tabling of the dichotomised qualitative data as described by BENNINGHAUS (1994)[572]. This tool provides the possibility to identify correlations between the variables, categories and the interviewees. Furthermore, the dimension reduction through factor analysis was conducted[573]. No

[568] MAYRING (2000: 11)

[569] According to BENNINGHAUS (1994: 30ff.)

[570] As described in BENNINGHAUS (1994: 92ff.)

[571] As described in BENNINGHAUS (1994: 168ff.)

[572] BENNINGHAUS (1994: 171ff.)

[573] As proposed by SCHNELL, HILL AND ESSER (1995: 152ff.)

significant results were established, however. In the case of ordinal variables and ranked data, as collected with *tables 21* and *23*, the data was tested with the SPEARMAN rank correlation[574].

5.4 Critical Annotations

Generally, the selected methods have proven to be adequate within the scope of the present research context. As the results chapter reveals, the major research questions could be answered and the research hypotheses could be tested. Problems occurred during the application of the snowball selection system and during the statistical analysis of the data.

5.4.1 Problems of the Snowball Selection Approach

The snowball selection approach provided the possibility to interview a large number of the manufacturers of bamboo products at the micro-level. However, it was not altogether apparent whether the actors interviewed represented the more socially and economically active members of the target population. It had to be assumed that actors entertaining strong private or economic contacts with a large number of other members of the target population were more likely to be named by them. This problem has already been recognised and described by SUDMAN AND KALTON[575] and GABLER (1992). To resolve this problem, they recommended enlarging the number of interviews in order to ensure a greater probability of new snowball contacts and intensifying the questioning for snowball contacts through repetition of the questions and changing the questioning style.

In the current context the aim of the snowball system was to approach as many actors as possible. The goal was to investigate all socio-economic compartments of the target population, e.g. the micro-enterprises, SMEs and large scale producers. The representativeness of the sample of these actors was only of secondary interest. The results show that actors from all socio-economic compartments were interviewed. However, the total size of the target population is still unknown, as is the representative percentage of the actors interviewed in the respective compartment. Due to the fact that the contacts of market competitors were asked for, and not social network contacts, which was the original application aim of the snowball system approach, the interviewees were probably better informed with respect to the location of these competitors and willing to name them.

---

[574] BENNINGHAUS (1994: 264ff.)

[575] SUDMAN AND KALTON (1986: 411ff.)

5.4.2 Statistical Analysis of Research Data

Due to the expected low number of interviews and the small quantity of data, it was clear that the statistical analysis of the data could only serve as a complementary element in the interpretation context. However, the body of quantitative micro-economic data and the quantified qualitative information of the meso-level interviews delivered a statistically researchable database, which deserved at least an attempt at statistical analysis. SILVERMAN[576] warns that such attempts at statistical analysis in qualitative research contexts can fail in terms of helping to find solutions to qualitative research problems and possibly be misleading in the interpretation of the data. Therefore, it was attempted to cautiously use the expected statistical analysis results. An effort was made to analyse the data by means of SPEARMAN'S rank correlation[577] and the PEARSON correlation coefficients[578]. Both analysis attempts did not result in any significant dimension reduction or in significant correlations between categories or variables. However, some trends were established, which could be used in the interpretation of the data. For the further statistical analysis of the data, simple methods such as cross tabling were preferred.

5.4.3 Consequences for Data Reliability and Interpretation

The aforementioned problems, which occurred during the research process, had to be considered while interpreting and validating the data collected and analysed. The non-representativeness of the micro-level entrepreneurs, in particular, must be emphasised in order to avoid premature generalisation of the research results. This fact was considered in both the presentation of the research results (*chapters 6 - 8*) and during the discussion in *chapter 9*. Only careful triangulation and testing of various possible interpretation contexts led to results for the micro-level, which could then be generalised for the whole sector. Part of this process was the use of the statistical analysis results, which could in fact only support the qualitative interpretation of the data. It was necessary to avoid handling the interpretation the other way around, and using the qualitative information to interpret the quantitative data.

[576] SILVERMAN (2001: 29ff.)

[577] BENNINGHAUS (1994: 264ff.)

[578] BENNINGHAUS (1994: 333ff.)

6. Research Results: Meso-level Analysis

The results of the meso-level analysis of the Colombian Guadua sector are based on data analyses of 43 structured guideline interviews conducted in summer 2002, as well as a set of open interviews conducted in early 2004 in the Colombian study region. Additional information was gathered from secondary data and literature research[579]. This chapter reveals the strategies and activities, as well as the patterns of interaction of the meso-level actors. Consequences of those interactions aimed at the promotion of the SMEs in the Guadua sector are highlighted and correlations are described. Furthermore, it will be emphasised what role the institutional embedding of the meso-level actors plays with regard to the actor's strategic capacities and their promotion activities within the Guadua sector. For the analysis of the meso-level the following research questions were defined:

- What are the meso-level actors' views about the present situation and the future development of the Guadua sector?
- How and why do they promote the Guadua sector? What are their strategies?
- How do the meso-level actors interact (co-operate/co-ordinate/compete) and what are the reasons for these interactions?
- Do the interactions between the meso-level actors determine their promotion activities?

6.1 Introduction to the Meso-level

In Colombia, during the last five years a number of regional and national actors have emphasised their interest in the improved use and marketing of bamboo products. Regional environmental authorities developed regulations for the use of the natural and planted bamboo resources in the study region. At the same time, educational and research institutions in the research region started to offer further education, such as construction and handicraft courses and lectures at universities. The increasing interest in the bamboo resources is reflected by the increasing number of scientific research publications on bamboo in Colombia. *Graphic 2* illustrates this increasing figure of research documents about Guadua related topics.

Since the first publications about *Guadua angustifolia*, the socio-economic importance of this resource for the regional actors in the *'Eje Cafetero'* has been mentioned. Before the background of the worsening socio-political and agro-economical conditions in the study region, the claims of the regional actors to the local bamboo resources have increased and led to intensified Guadua related activities. It is assumed that the bamboo resources contribute to the stabilisation of the rural household economy and that the industrial processing and the

[579] See *chapter 5* on the methodical details and *chapter 4* for the description of the study region.

national and international marketing of the bamboo products will generate employment and income for the unemployed and the increasing numbers of migrants[580].

Furthermore, in January 1999 an earthquake devastated vast areas of the departmental capital Armenia and the surrounding municipalities. The earthquake left concrete and brick buildings in ruins, whereas traditionally built Guadua based constructions[581] remained intact. The German technical co-operation agency GTZ in co-operation with local entities organised a fast track programme to build Guadua houses for the homeless victims of the earthquake. In the course of this programme, three villages have been reconstructed. The recognition of the excellent anti-seismic characteristics of Guadua buildings and the usage of Guadua for the construction of social housing demonstrated the local population and actors the possibilities, which the utilisation of the Guadua resources offer.

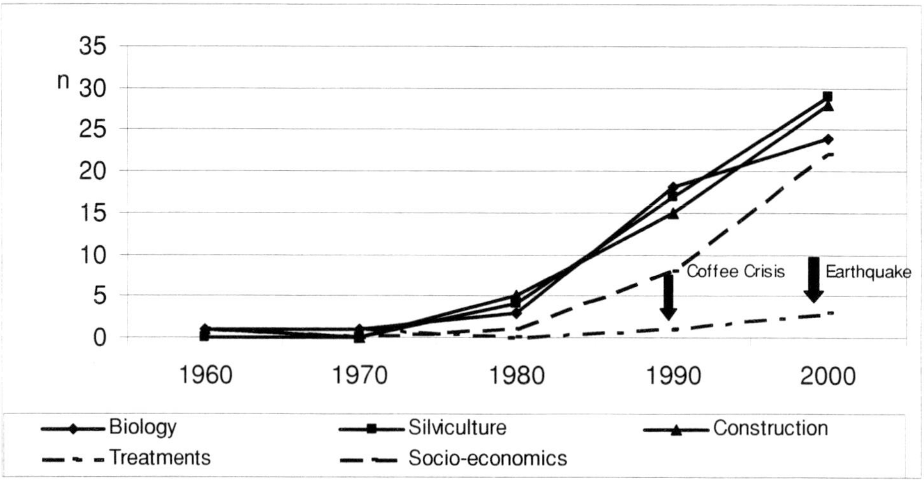

Graphic 2. Numbers and topics presented of Guadua related publications in Colombia from 1950 to 2000 (based on ARBELAÉZ, 2001)

The role of the Guadua resources for the local population and its economic importance for the rural households and small-scale enterprises was first mentioned by MEJIA in the year 1952. From this year on the topic of the contribution of Guadua and Guadua products to rural livelihoods and the regional economy was taken up several times by a number of authors

[580] OSPINA (2000)

[581] The architectural style is called '*Bareque*'. For detailed information on constructions with Guadua, see HIDALGO 2003.

from various backgrounds, especially during the 1990s[582]. These publications about the role that the Guadua resources play in the Colombian society and economy were accompanied by increasing research about the feasibility of the production and manufacturing processes, as well as the national and international marketing potentials of Guadua products[583]. The most important books about *Guadua angustifolia*, which cover the whole width of aspects related to the topic from the seed/clone to the final manufacturing processes, have been published in the past ten years[584].

6.2 Meso-level Guideline Interviews 2002

In the course of this study 43 meso-level actors were interviewed[585]. Due to different institutional backgrounds the actors were subdivided into different groups in order to classify their activities and interactions according to the actor-based institutionalism approach described in *chapter 3.3*. The institutional backgrounds determine the actors' obligations and duties, as well as their level of influence and rights to participate in decision-making processes. *Table 28* provides an overview of the interviewed actors and their institutional backgrounds, a detailed list of the individual actors can be found in *annex 3*.

Table 28. Organisational backgrounds and numbers of interviewed meso-level actors

Actor Group	Number
Governmental Organisations	13
NGOs	4
Construction Sector Assoc.	5
Agricultural Producer Orgs.	3
Research Organisations	7
Combined public-private Actors	5
Independent Experts	6

[582] E.g.: RINCÓN (1977); CORTES AND MONTOYA (1986); BELTRAN AND QUIMBAYA (1987); POSADA (1991); GONZÁLES AND POSADA (1996); GIRALDO AND SABOGAL (1999: 21ff.) and BUCHELLY (2001)

[583] For example: PEREZ (1980); CHACON AND ZULUAGA (1990); BELTRÁN AND PASTRANA (1991); LÓPEZ (1999); LÓPEZ AND SIERRA (2000); LÓPEZ (2001); RUIZ AND AMELINES (2002); STAMM (2002); and REPUBLICA DE COLOMBIA. MINISTERIO DE COMERCIO EXTERIOR (2001).

[584] These studies are: RÍOS (1994) ; GIRALDO AND SABOGAL (1999); JUDZIEWICZ ET AL. (1999); MORAN (2001); HIDALGO (2003) and CASTAÑO AND MORENO (2004)

[585] For the purpose of this research, the total of the Guadua-sector related meso-level actors was sought to be included in the research process. The actors could be identified with the help of lists which named the participants of the most important Guadua related activities and events held in the study region in the year 2001. See chapter 5.

6.2.1 Meso-level Actors` Perception of the Guadua Sector

The 43 structured guideline interviews were recorded, transcribed and analysed according to the principles of qualitative data analysis[586]. The analysis resulted in the creation of eight answer categories and 39 answer items derived from the set of guideline questions. The questions dealt with problems and obstacles to the development and promotion of the Guadua sector[587]. One answer item is identical to one statement/answer to a guideline question and was defined when it was mentioned by at least one interviewee. Cognitional items constitute one category. In *annex 4* these categories are listed and assigned to the respective analytical level of the concept of systemic competitiveness.[588]

Graphic 3 presents those items named by more then 33% of the interviewees. The figure reveals that for all interviewed actors the items of the categories '*inter-organisational problems*', '*deficits in interaction and involvement*' and '*general problems of the sector*' are the most important ones. The leading item '*lack of valuation of the bamboo resources and products*' (75%) belongs to the category '*Socio-cultural problems*'. According to the meso-level actors the most important obstacles to the development and the promotion of the Guadua sector are shortcomings of co-ordination, co-operation, as well as communication and organisation. Furthermore, the lack of public and private investments in the sector (40% and 56%) is seen as a hindrance to development.

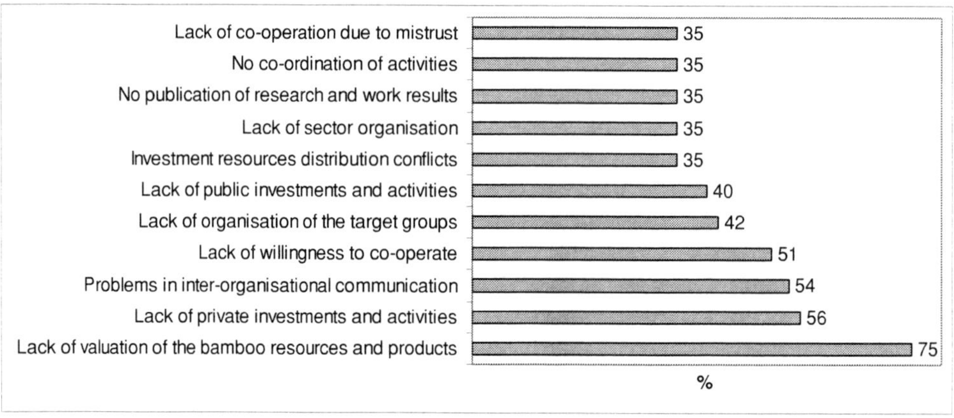

Graphic 3. Items named by more than 33% of the interviewees (n=42)

[586] See *chapter 5.3*

[587] See *annex 2, questionnaire 1*

[588] Compare *chapter 3.1*

Analysing the different actor groups, the named problems and obstacles to the Guadua sector's development showed rudimental patterns of distribution within and among these actor groups. *Table 29* presents three sets of obstacles and problems and the correlating groups of actors. These three sets are based on those items, which were frequently named in combination with each other, e.g., the interviewees who mentioned institutional problems and obstacles to the sector's development also named socio-cultural obstacles and problems. When the majority of group members (more than 2/3 of the actors in an actor group) answered according to one of the three patterns of answering, the answers were assigned to the respective set of items. To identify the three sets of cognitional items, statistical factor correlation was applied. However, there were no significant correlations. Thus, no statistical analysis is presented. Nevertheless, correlation tendencies could be identified, which are cautiously interpreted in *table 29*.

The set of institutional and socio-cultural problems was predominantly named by governmental actors and researchers, as well as by the public-private actors. The set of inter-organisational problems was emphasised by the NGOs, the group of independent experts and the associations of the construction sector. Only for the organisations of agricultural producers the decision-making processes and lacking financial resources were the most important problems.

Table 29. Related categories of problems and obstacles according to actor groups

Actor Group/ Set of obstacles	Institutional and socio-cultural obstacles	Inter-organisational problems and lacking organisation of the sector	Deficiencies in the decision-making process and intra-organisational problems
Governmental Organisations	*X*		
NGOs		*X*	
Associations of the Construction Sector		X	
Agricultural Producer Orgs.			*X*
Research Institutions	*X*		
Independent Experts		X	
Public-private Actors	X		

6.2.2 Meso-level Actors` Influencing Strategies

To analyse the interviewed actors' possibilities to influence the development and promotion of the Guadua sector, the interviewees were asked about the influencing strategies of their organisations[589]. *Table 30* reveals that especially the governmental actors and the associations of the construction sector apply a wide range of strategies to gain influence, including drafting bills for future laws and regulations. The other groups of actors are less active since they mainly focus on their roles as consultants and researchers. Actors of all groups use informal personal contacts as instrument to influence Guadua sector related decision making.

Table 30. Strategies of actor groups to gain influence on the Guadua sector

	Position of Decision-maker	Participation in the decision-making process	Consulting decision-makers	Publication of research results	To convince with previous project results	Informal personal contacts	Drafting bills
Governm. Orgs. (n=13)	X	X	X	X	X	X	X
NGOs (n=4)					X	X	
Construction Sector Assoc. (n=5)		X	X		X	X	X
Agric. Producer Orgs. (n=3)			X	X	X	X	
Research Inst. (n=7)			X	X	X	X	
Indep. Experts (n=6)		X	X	X		X	
Public-private Actors (n=5)		X			X	X	

589 A strategy of gaining influence was assigned to the whole group of actors, if more than 2/3 of the group's interviewees mentioned this strategy during the guideline interview.

6.2.3 Meso-level Actors` Motivation to engage in the Guadua Sector

During the interviews the actors were asked to give the reasons for their involvement in the Guadua sector and since when they are involved in Guadua related activities. The basic development goals of the interviewed actors for the Guadua sector were surveyed. Two main developmental goals and strategies for the Guadua sector could be identified:

- Regional development promoted by the industrialisation of the Guadua sector, especially considering economic and social aspects with a focus on the international market. This was mainly a goal of the governmental actors and partly the agricultural producer organisations, as well as the public and private actor organisations.
- Stabilisation of the agricultural society through taking into account ecological and social aspects and focusing on regional strengths and weaknesses in a national context. Researchers, actors of the construction sector, agricultural producer organisations and NGOs emphasised this goal.

The years in which each of the 43 interviewees began getting involved in the Guadua sector are shown in *graphic 4*. The majority (more than 50%) started their activities after 1998. The earliest date is 1980. However, it is very likely that other actors had begun such activities in the past and abandoned them meanwhile. These activities could not be assessed within the scope of this study. Only actors who are still actively involved could be considered.

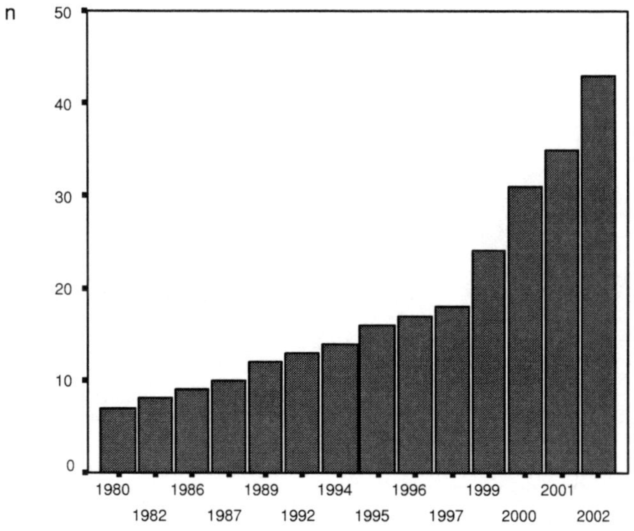

Graphic 4. Culminating number of meso-level actors engaged in Guadua related activities since 1980 (n=43)

When asked why they started to engage in Guadua related activities, the interviewed actors agreed that the Guadua resources offer the possibility to improve the economic and social situation in the study region. The involvement of certain actors is closely related to their institutional background. *Table 31* provides an overview of the reasons given by the actors of the different groups for their involvement in Guadua related activities.

Table 31. Reasons for the involvement of meso-level actors in Guadua related activities

Group of actors	Institutional legitimisation	Set of reasons for involvement in the Guadua sector
Governmental Organisations	Their duties and rights by law and by their interpretation of normative role-understanding	To protect the Guadua resources as part of the environment To promote the economic and social well being through production of and trade with Colombian natural, agricultural and industrial products
NGOs	Their mission statement and their moral principles	The Guadua is a means to improve the livelihood of the people The Guadua is a means to combine sustainable use of natural resources with meeting the local population's needs
Construction Sector Associations	As private organisations of market participants	To ensure the supply of their members with Guadua raw materials To ensure qualitative standards of Guadua products and constructions To improve access to markets In general to improve the competitiveness of their members
Agricultural Producer Organisations	As private organisations of agricultural producers	To ensure the livelihood and well being of their members To secure the production and sale of agricultural products including Guadua Guadua has the potential to contribute to agricultural production systems and improve their competitiveness
Research Institutes	As scientific organisations	To do research on open scientific questions about Guadua To educate people about Guadua To contribute to the sustainable use of the resources and to the social and economic well-being of Colombia through education To act as service providers for private and public clients
Independent Experts	As individuals	Personal interest and satisfaction Passion for the Guadua plant and its uses Professional interest in the solution of Guadua related problems A source of income
Public-private Actors	As public-private actors according to their legal duties and as service providers for their members	The Guadua offers the possibility to improve the social and economic well-being of their interest groups Contribution to growth and development

6.3 Meso-level Questionnaire Interviews 2002

During the interviews, the actors were asked to choose from a list of promotion activities in order to describe and rank their strategies to achieve the development goals with respect to the Guadua sector. The listed promotion activities comprised the activities as listed in *table 32*.[590] In case the listed options did not include an activity pursued by the interviewed actor a new category could be added. The interviewees were asked to rank the promotion and development activities of their organisations on a scale from '1' (for lowest intensity in the efforts in the named activity) to '5' (for highest possible degree of intensity in employed efforts)[591]. In doing so, the interviewees compared the activities of the years until 1997 with the period from 1998 to 2002.

Table 32. Categories of promotion activities

Category	Activities
Research and Development (R&D)	Basic biological research; Silviculture research; Market research; Product research and development; Process research and development
Expansion, Dissemination and Education (E&E)	Entrepreneurial extension; Technical extension; Professional education; PR
Institutions-building and Initiation of Projects (I&I)	Participation in Guadua related associations; Participation in institutions-building; Funding of bamboo projects

6.3.1 Activity Mixes of Meso-level Actor Groups

The *net-diagrams 1-6* illustrate the mix of activities of each actor groups and the changes, which took place between the time until 1997 and the period from 1998 to 2002. These diagrams reflect the average ranking, which each activity received by the different groups of actors. The average ranking per group of actors per activity was calculated (sum of individual rankings for one activity/ number of corresponding interviewed actors of one group). The possible maximum result is '5' for the most important activity. The lowest result is '0' for no engagement at all. Hence, for each group of actors a net-diagram could be created, which shows the average degree of importance of the activities within an actor group.

The group of independent experts is not included in this analysis. These experts are usually employed and paid by the other actors. Therefore, their activities are already included in those

590 See *annex 2*, *questionnaire 1*, and the methods *chapter 5*.

591 Compare *chapter 5* and *questionnaire 1 (Annex 2)*

actors` rankings. To avoid presenting the same or similar results twice, the experts were not included in the individual ranking analysis.

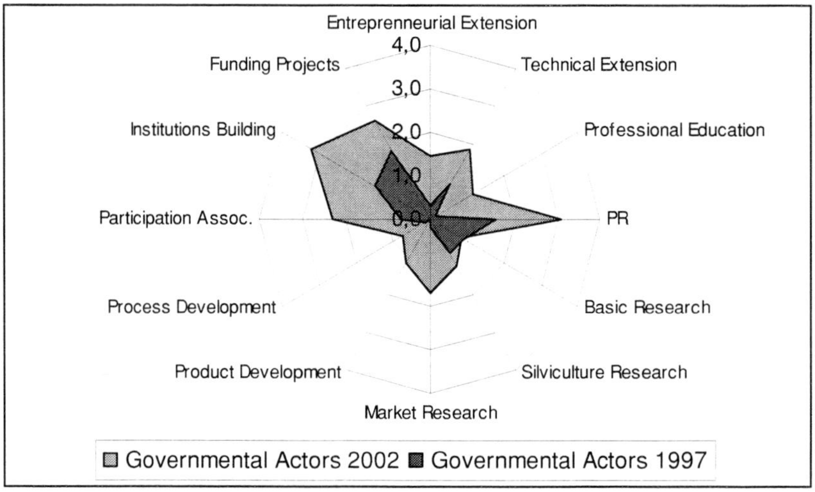

Net-diagram 1. Governmental actors' activity mix until 1997 and between 1998 and 2002

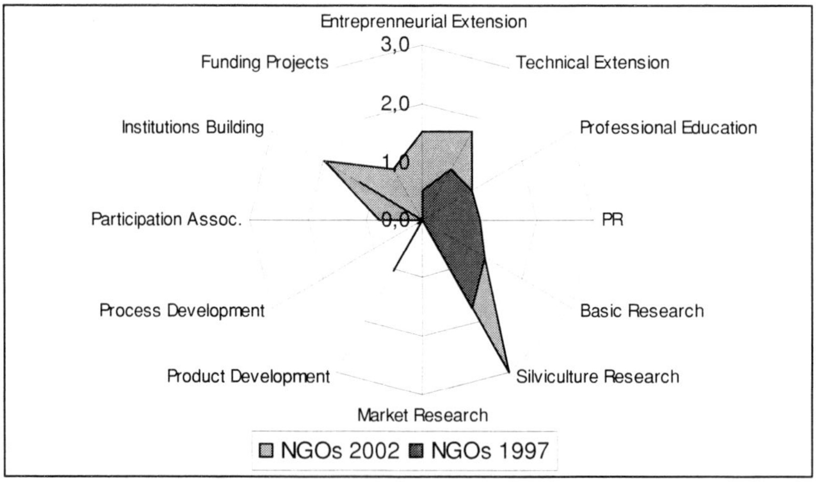

Net-diagram 2. NGOs' activity mix until 1997 and between 1998 and 2002

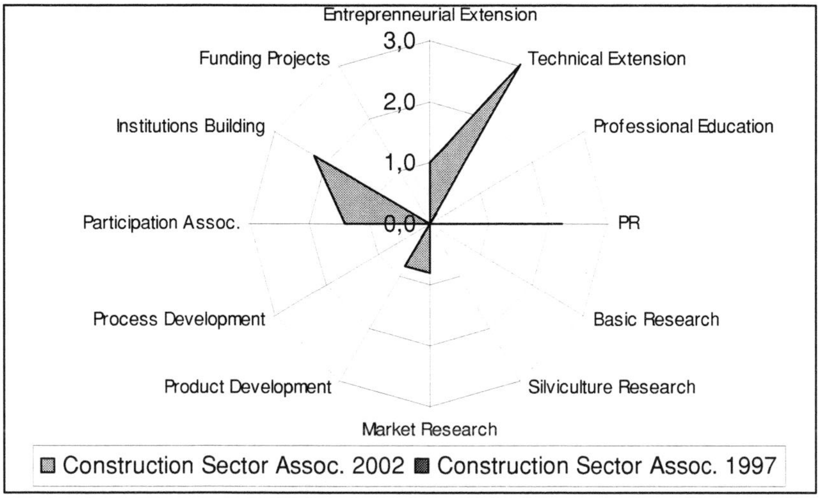

Net-diagram 3. Construction sector associations' activity mix until 1997 and between 1998 and 2002

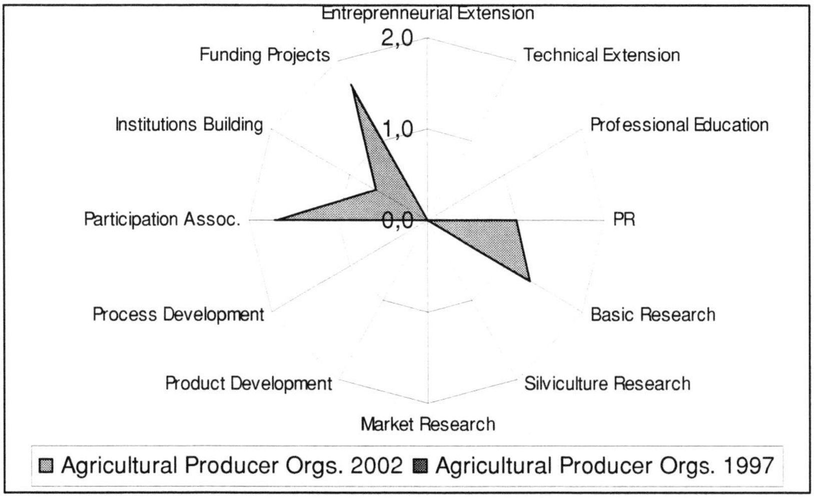

Net-diagram 4. Agricultural producer organisations' activity mix until 1997 and between 1998 and 2002

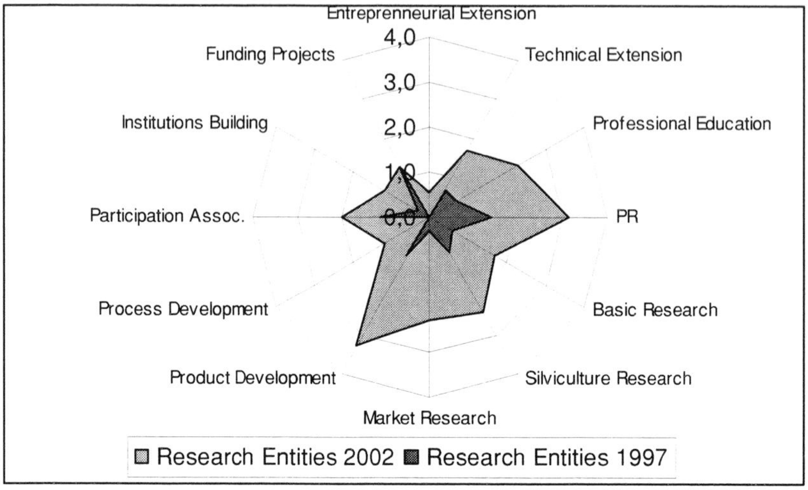

Net-diagram 5. Research entities' activity mix until 1997 and between 1998 and 2002

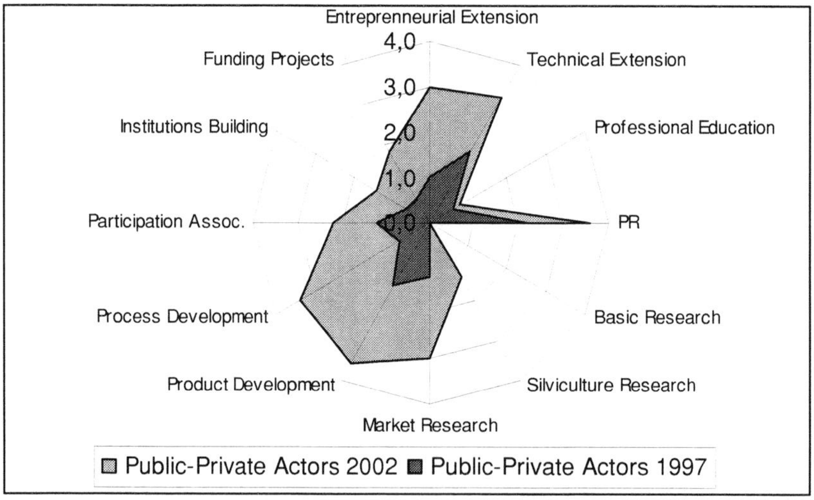

Net-diagram 6. Public-private actors' activity mix until 1997 and between 1998 and 2002

The net-diagrams reveal the different groups' foci and the changes of their basic strategies. Research organisations, as well as the public-private actors are mostly active in R&D. In the period from 1998 to 2002 they increased their efforts with regard to Guadua related activities significantly. The research organisations intensified their research activities whereas the public and private actors put more emphasis on the development of the procedures and the products. Within the sector these two groups of actors are also intensely engaged in the creation of institutions and organisational processes (I&I), as well as extension, education and dissemination activities (E&E). As in the case of R&D-activities, they increased their efforts in these fields of activity in the period from 1998 to 2002, too.

The governmental organisations and the NGOs show moderate engagement in R&D and E&E activities. The governmental actors' involvement in I&I activities is higher than that of the other actor groups. In contrast, the NGOs together with the associations of the construction sector show the lowest level of I&I activities. Both groups of actors increased their Guadua related activities in the period from 1998 to 2002. The governmental actors account for the greatest rise in I&I and E&E. The NGOs particularly expanded their activities in the field of I&I activities.

The associations of the construction sector and the agricultural producer organisations are hardly engaged in Guadua related R&D. However, in contrast to the time until 1997, during which they undertook nearly no Guadua related activities, they started some research projects. At the same time, actors of the construction sector have always disseminated their experiences made with Guadua constructions and construction guidelines (see PR activities shown in *net-diagram 3*). Due to no activities recorded in the neighbouring compartments of the net-diagram visualisation, the graphic (*net-diagram 3*) fails in this case. However, this is the only time the problem occurred. The main focus of the construction sector associations is on E&E activities. The agricultural producer associations continue to be strongly engaged in I&I. Before 1998, the former were virtually not involved in I&I activities, whereas the latter were already active in that field.

6.3.2 Portfolio Analysis of Actor Groups` Activity Mixes

In figure 5, all the six groups of actors were set into relation to each other using a triangular portfolio presentation of their activity mixes with the three corners representing R&D, E&E and I&I activities. The averages of the named activities in R&D, E&E and I&I were calculated for each group of actors (sum of individual rankings in one field of activities/ number of corresponding interviewed actors of one group). The possible maximum result is '5' for the highest level of engagement in a field of activity, whereas the lowest result is '0' for no engagement at all. For each group of actors a pie chart could be created, which shows the average distribution of the three areas of activities within an actor group. *Figure 11* illustrates which colour of the pie chart represents which area of activity. In *figure 12*, the pie charts, representing the six actor groups, are placed into the triangle according to the level of importance of the three areas of activities within that group.

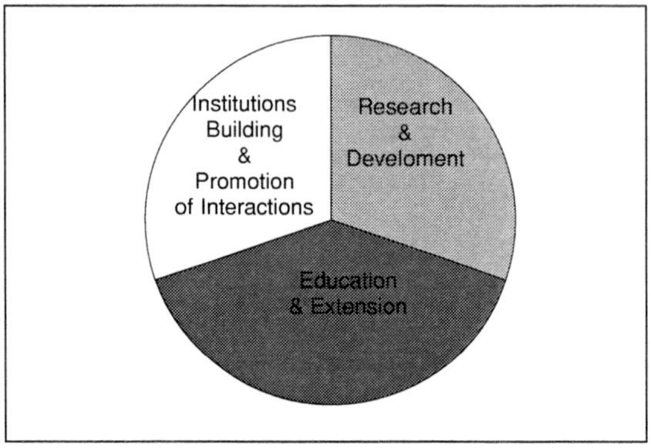

Figure 11. Colours representing the areas of activities

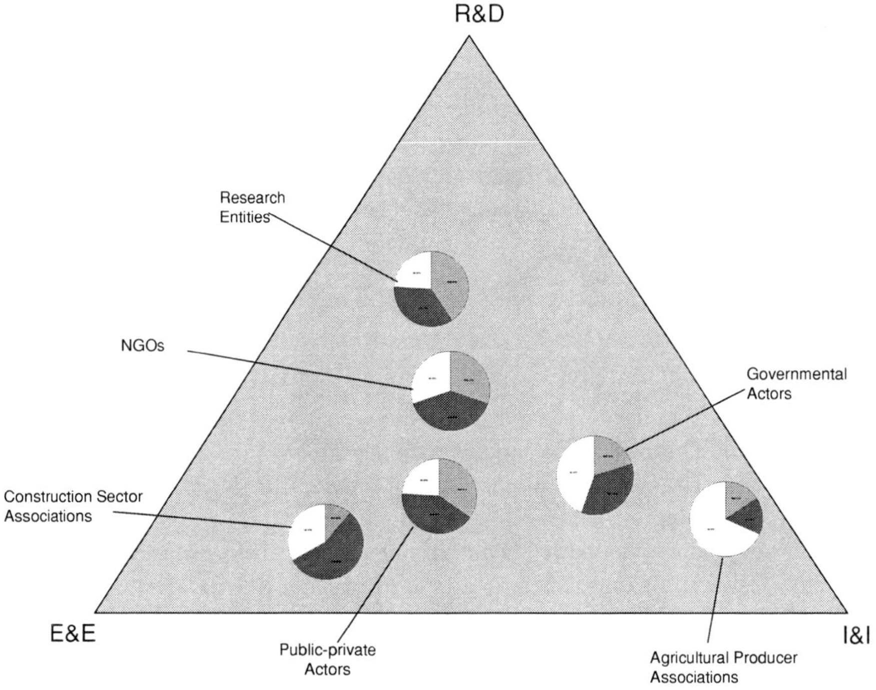

Figure 12. Position of actor groups within the triangle of R&D, E&E and I&I according to promotion activities conducted from 1998 to 2002

Figure 12 highlights that the actors try to achieve a mix of activities according to their given resources. The major differences of strategies either apply to their engagement in I&I or R&D activities. The agricultural producer organisations emphasise the institutional work and the facilitation of communication among their micro-level members. The governmental actors mainly engage in E&E and I&I activities. Research organisations dominate the R&D activities. NGOs and the public and private actors are also involved in this area. The NGOs mainly focus on silviculture research, the research organisations conduct basic and silviculture research, whereas the public-private actors emphasise research on the development of products and processes. The construction sector's actors focus on technical development and educational measures, they represent the group of actors with the lowest intensity in R&D activities.

6.3.3 Patterns of Meso-level Actors` Interactions

To assess the ability to co-operate with each other and to co-ordinate activities the interviewed actors were asked to name their co-operation partners in the field of Guadua sector promotion and development. They were supposed to describe the objectives of these co-operations, which could focus on either R&D, E&E or I&I. The actors had to explain the links with their co-operation partners with regard to technology, human resources, institutional and financial aspects[592]. The independent experts were excluded from this analysis, because their answers to the questions about co-operation usually referred to their employers. Hence, their understanding of co-operation greatly differed from that of all the other actors. In this chapter, only the answers of the complex (organisational) actors[593] are presented.

Graphic 6 shows the average number of co-operations per actor per actor group in the three fields of activities. In total, 126 co-operations were named by the 37 interviewed actors. This figure includes multiple nominations since one co-operation could be mentioned by all the participating partners. In case two or more co-operation partners belonged to the group of the 37 interviewed actors the co-operations were counted accordingly. Since the aim of this assessment was to describe each actor's degree of interaction the multiple nominations do not lead to an exaggeration of the co-operative activities, but reflect the rate of involvement in co-operations per actor and group respectively. *Graphic 5* illustrates that most co-operations are in the area of R&D, followed by I&I and E&E related co-operations.

In this analysis, the group of public-private actors is the group with the highest average number of co-operations per actor. However, all actors of all groups are involved in at least two co-operations. The public and private actors mainly co-operate in R&D, so do the NGOs and the agricultural producer associations. The governmental actors put emphasis on co-operations in the field of I&I. The research organisations achieve comparable levels of co-

[592] See *chapter 3.3* and *annex 2*

[593] The organisational actors are the complex actors as described in *chapter 3.3*.

operation in R&D, as well as E&E, but co-operate less in I&I activities. The associations of the construction sector show similar tendencies. No group of actors stands out with regard to E&E co-operations.

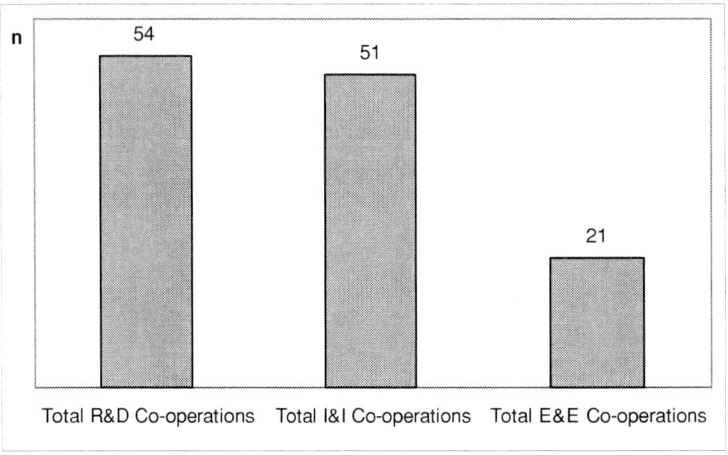

Graphic 5. Total number of co-operations named by all meso-level actors. Subdivided according to fields of activity (n=126)

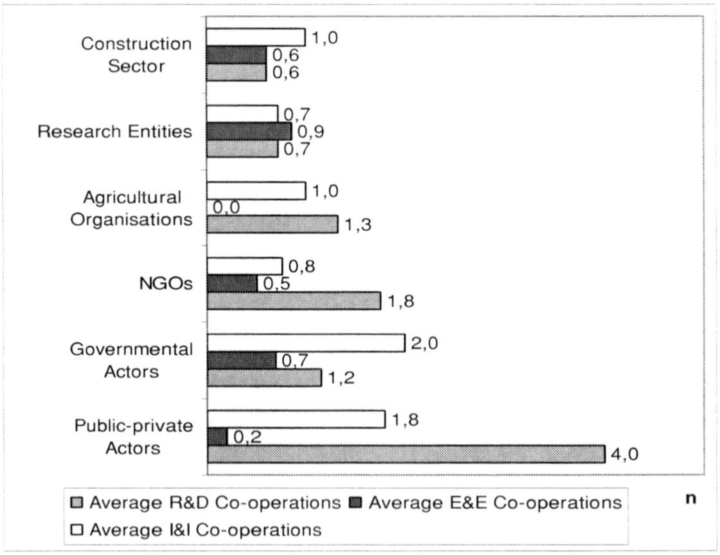

Graphic 6. Average number of co-operations per actor per group. Subdivided according to co-operation activities

The interview analysis revealed 126 co-operations entertained by 51 actors. These 51 actors comprised the 37 interviewed actors, who accounted for 112 co-operations and 14 additional actors, who were engaged in 14 co-operations. In *graphic 7*, the co-operations entertained by these 14 actors are labelled 'others'. This group consists of one university, two Colombian NGOs, two banks, the Colombian Bamboo Society or one of its members and the national Colombian research funding organisation COLCIENCIAS. The Colombian Bamboo Society was mentioned as co-operation partner six times[594] and COLCIENCIAS was named three times. The graphic shows the high number of co-operations with governmental actors. From the group of 37 interviewed actors, the 13 interviewed governmental actors were named 66 times as partners in co-operations, the five public-private actors were named 23 times, the seven research entities 15 times, the three agricultural producer associations four times, the four NGOs three times and only once one of the five interviewed associations of the construction sector was named as co-operation partner.

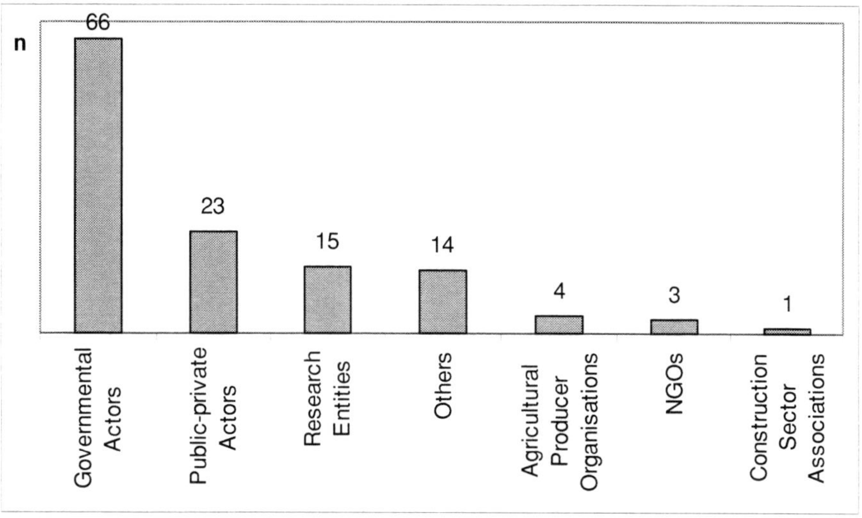

Graphic 7. Co-operation partners of meso-level actors (n=126)

The co-operations of the interviewed actors are illustrated in *venn-diagram 1*, which was developed on the basis of a cross table analysis (see *annex 5*) of the interviewed actor groups and the mentioned co-operation partners The exact number of corresponding interactions is noted at the side of the line close to the respective actor group. The width of the line

[594] For a more detailed description of the Colombian Bamboo Society see *chapter 6.3.5*.

represents the number of co-operations among the different groups of actors. The size of the circle of each group of actors represents the number of co-operations between the actors within that group (the figures in the brackets present the number of co-operations with actors within the same group of actors). The co-operations between the governmental actors and the public-private actors, as well as the co-operations between the governmental actors and the research institutions dominate the presented network.

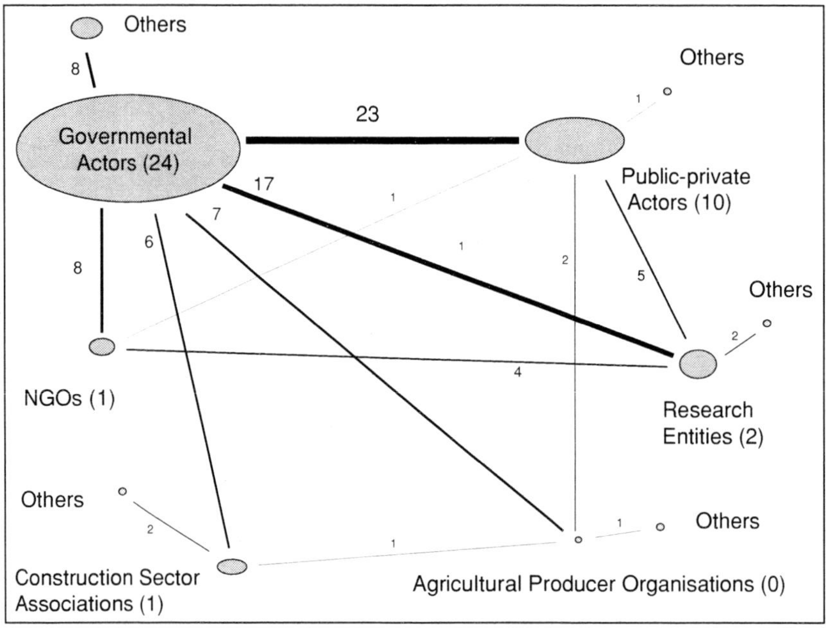

Venn-diagram 1. Network of co-operations between the meso-level actors

6.3.4 Correlations between Co-operations, Promotion Activities and Actors` Self-evaluation

In order to test the effects of co-operations on the actors' strategic capacity and their promotion activities, correlations between the co-operations, the mix of activities and the self-evaluation of the meso-level actors were sought. The interviewees were asked to evaluate their activities carried out so far and to compare their activities until the year 1997 with those of the period from 1998 to 2002. The self-evaluation was ranked from '1' to '3': '1' meant 'poor results of the conducted activity', '2' was 'satisfying' and '3' stood for 'very good results'. The evaluation should take into account the resources and efforts employed to conduct the respective activity[595]. In *annex 6* the results of this self-evaluation are presented.

The results of the meso-level actors` self-evaluation reveal that the governmental actors, the agricultural producer organisations, the construction sector associations and the public-private actors associations tend to mark all of their activities between 1998 and 2002 more positively. The NGOs and the research institutions consider their recent activities to be worse than during the period until 1997 or equal.

The correlation analysis was done according to SPEARMAN since the variables are ordinal (ranks) and non-Gaussian (small number of research totals and group of actors sub totals)[596]. The following variables were correlated: the average ranking of R&D, E&E and I&I activities per actor per group of actors done in 2002; the average self-evaluation of each actor and each group of actors with regard to all three fields of activities done in 2002; the average number of co-operations per actor per actor group in all three areas of activities. Due to the small number of cases (six cases for the six considered groups of actors), the results of the correlation analysis do not withstand statistical requirements. However, they help to interpret the research data and identify tendencies of the interrelations between the variables. The complete tables on which the correlation analysis is based can be found in *annex 6*.

The analysis of interrelations between the ranking of activities, the self-evaluation and the mean number of co-operations showed that actors, who are very active in R&D and E&E, tend to evaluate their activities "successful". On the other hand, strong involvement in I&I activities and the self-evaluation were characterised by a high negative correlation factor. Thus, actors tended to classify these activities as 'unsuccessful' although they were engaged in a high number of them.

The analysis of correlations between R&D, E&E and I&I revealed a negative correlation between R&D co-operations and E&E co-operations. This indicates that a high number of co-operations per actor in the field of R&D is correlated to a low number of co-operations per actor in the field of E&E activities. A positive correlation was found between the degree of involvement in E&E and I&I activities, i.e. actors who are very active in one of the two fields are also very active in the other[597].

6.3.5 Independent Experts Analysis

As mentioned in the preceding chapters, due to problems with regard to the comparability, the results for the group of interviewed independent experts are presented separately. The independent experts form a group of meso-level actors, which consists of individuals working independently in the field of Guadua sector development and promotion. The role they play at

---

[595] See *chapter 5* and *annex 2, questionnaire* 1.

[596] See BÜHL AND ZÖFEL (2000: 322f.)

[597] This is the only significant correlation according to the Spearman Rank Correlation (*Annex 7*).

the meso-level is characterised by two aspects: 1) They are formally employed by changing meso-level actors to conduct research, to give lectures, to participate in conferences and seminars, to give lessons in extension campaigns, etc. In these cases, the activities were assigned to the respective organised meso-level actors, for which the results were presented in the previous chapters; 2) The independent experts conduct private research and experiments, publish their results, write books and maintain informal relations not only with the Guadua sector's meso-, macro-and micro-level actors, but also with actors from other economic sectors, such as the forestry sector and the wood processing industry.

For the purpose of this study six of these experts were interviewed during the field campaign in 2002 and a seventh one was consulted in the year 2004. Their expert status was defined based upon a set of criteria. These criteria referred to the period of time the individual is engaged in Guadua related activities, the number of scientific publications and the reputation of the expert amongst the other actors of the Guadua sector. Since these criteria are not objectively revisable and the quality of individual knowledge and expertise may vary, the results of these expert interviews have to be presented and interpreted carefully. However, the description of these groups' activities and perceptions is a useful source of information and a necessary point of view to complete the analysis of the Guadua sector in this case study. The majority of the interviewed experts (six out of seven) have been working in the field of Guadua related research and other activities since the early 1980s[598]. Five of the seven interviewed experts are leading members of the Colombian Bamboo Society (SCB[599]).

The actors\` basic understanding of the potential contribution of bamboo to regional development processes is embedded in a social and ecological framework, thereby, the industrialisation process is perceived as the most desirable path to follow. The actors of this group draw a large share of their scientific and professional justification and identity from the bamboo topic. Besides the personal aspects, the Guadua topic presents a source of income to this group of actors. Due to the current boom they feel affirmed and, at least on a regional scale, they are recognised and acknowledged.

The experts usually interact with the other groups of actors in Guadua related research and promotion activities, and with respect to Guadua related questions and decisions they consult organisations of all groups of actors. They are often called upon as speakers at conferences and seminars and they work as hired instructors and lecturers at universities and in extension activities. The experts stated, that as independent researchers and professionals they usually lack the financial resources to pursue their independent work. Their involvement in the decision-making process depends on informal personal contacts. However, these contacts may reach the highest level of Colombia's political administration.

---

598 Compare *graphic 4* in *chapter 6.2.3*.

599 Spanish: *SCB=Sociedad Colombiana de Bambú*

The independent experts identified the following problems of the bamboo sector: deficiencies in inter-organisational communication, the general lack of willingness to co-operate, and the lack of public and private investments, as well as the lack of public and private interaction and co-operation. According to these experts there are general deficits in the institutional framework to entrepreneurs, foreign investments and export activities. To them the lacking national security problem must be resolved in order to improve the general economic environment, which in turn will attract foreign investments and boost the bamboo sector's development.

The experts perceive bamboo flooring and carbon fixation as the most promising bamboo products. Additionally, they would like to see more use of bamboo as construction material in the regional construction sector and social housing projects. According to their perception, deficits in biological and technical knowledge must be overcome, and profound knowledge about the international markets for the potential products must be acquired.

The experts are involved in a number of private activities, which they mainly fund themselves. In *table 33*, the various private activities of the independent experts are listed. Moreover, the table illustrates which informal personal contacts the experts have with actors of other groups and sectors to influence the development of the Guadua sector. The formal employment and service contracts, and the informal contacts and co-operations between the independent experts and other actors are of special interest to the sector since the experts account for a large pool of special tacit knowledge on the Guadua resources. They have individual channels to acquire knowledge from other sectors than the Guadua sector. This knowledge is imported into the Guadua sector and made available to the other actors and entrepreneurs through active dissemination or passive spill-overs.

Table 33. Activities and informal contacts of independent meso-level experts

Expert	Activities	Informal Contacts
1	• Furniture, laminated bamboo, boards, flooring, preservation, drying, etc. • Book publisher	• Wood processing sector • Universities
2	• Silviculture, propagation • Book co-author and reviewer	• Universities • CARs
3	• Furniture, laminated bamboo, construction • Book publisher	• Construction sector • Universities • Wood processing sector
4	• Preservation, drying	• Guadua sector micro-level enterprises
5	• Preservation, drying	• Guadua sector micro-level enterprises
6	• Institutional co-ordination	• Universities
7	• Silviculture, propagation • Book contributor	• CARs • Forestry sector

6.4 Complementary Meso-level Actor Interviews 2004

Besides the actors who actively and intentionally influence the Guadua sector's development, there are others who influence the sector unintentionally or even without knowing that they do. Furthermore, various actors were not interviewed in 2002 since they were not openly involved in Guadua related activities. Thus, they could not be identified in advance. A number of these actors were interviewed during the second interview campaign in the year 2004 in order to analyse their impact on the Guadua sector. *Table 34* provides an overview of the number of interviewed meso-level actors in 2004 and their institutional backgrounds (the individual interview partners are listed in *annex 7*). The interviews were analysed separately and the results were used to help with the interpretation of the meso- and micro-level analysis (*chapters 7* and *8*). Some results could be organised systematically and are presented in the following sections.

Table 34. Number and organisational backgrounds of additionally interviewed meso-level actors in 2004

Organisational background	Number
Regional SME promoting organisations	3
Actors promoting R&D at national level	2
National Guadua related organisations[600]	2
Banks with SME credit lines	3
NGOs with SME credit lines	2
National organisations of the forestry and wood processing sectors	2

6.4.1 Representatives of the Colombian National Innovation System

In Colombia, like in many other countries, a National Innovation System (NIS) is established[601]. The NIS includes a number of organisations, which are interrelated and which are mainly co-operating in certain areas of R&D and partly in activities related to E&E and I&I. The participating actors in the NIS are usually governmental organisations (e.g. the ministry of trade and industry), state extension and education facilities (in Colombia, this is the SENA[602]), public research organisations (generally universities), public-private actors, such as chambers of commerce and trade and associations/organisations of economic sectors,

[600] These two organisations were founded in the years 2003 and 2004. Hence, they were not included in the 2002 research campaign.

[601] For a theoretical introduction into regional and national innovation systems see *chapter 3.2.3*.

[602] Spanish: *SENA=Servicio Nacional de Aprendizaje*

such as agricultural producer organisations or associations of the industrial sector. In Colombia, the NIS pursues three major strategies[603]:

- A network of regional R&D centres, which includes public universities and other governmental and semi-governmental organisations, as well as several research facilities of agricultural producer organisations and the industrial sector,
- The promotion of industries and SMEs through public-private actors and the provision of technical and entrepreneurial expansion, as well as education services,
- The provision of funding and credit schemes for entrepreneurs.

Actors of the NIS, who are actively involved in promotion activities of the Guadua sector and/or are present in the study region, were interviewed in 2002. All of those actors belonged to the groups of actors, which the first strategy applies to. The results of that interview campaign were presented in the preceding chapters. In the 2004 research campaign, actors who belong to the second and third categories were identified and interviewed.

In Colombia, 0.2% of the GDP was spent on R&D activities in 2001. Around 50% of the total were private investments, roughly 40% are investments made by research institutions and only 10% were state investments[604]. Over the last years there were almost no direct foreign investments in R&D activities. The main source for financing R&D in Colombia is the distribution of state-run funds through the national organisation COLCIENCIAS. Until 2002, COLCIENCIAS[605], which is the roof organisation of the Colombian NIS, had provided almost no funds for R&D projects in the Guadua sector. In 2004 the situation changed significantly and COLCIENCIAS has become one of the most important promoting organisations of the Guadua sector.

According to COLCIENCIAS staff, the little involvement in Guadua related research was due to the low number of research project applications in this field. However, since the year 2000 the number of research applications rose and consequently COLCIENCIAS increasingly supports Guadua related research. Before 2001, only one Guadua related research proposal was found in the COLCIENCIAS database (1996). That proposal was rejected.

---

[603] The Colombian NIS is presenting itself and its members in detail in the internet: www.colciencias.gov.co:8888/sni/index.php3.

[604] According to UNESCO database analysis (www.uis.unesco.org/TEMPLATE/html/Exceltables/science/GERDMay2004.xls) in 2004.

[605] Spanish: *COLCIENCIAS=Instituto Colombiano para el Desarrollo de la Ciencia y Tecnologia*

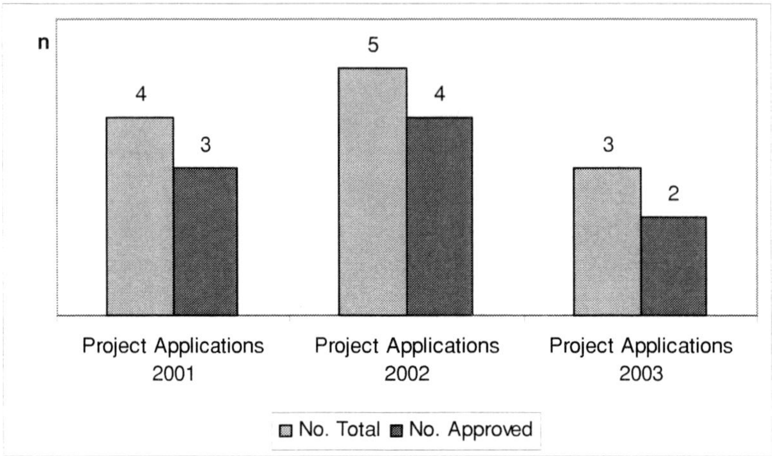

Graphic 8. Guadua related project applications and approved Guadua related projects by COLCIENCIAS from 2001 to 2003

Graphic 8 illustrates that more than 2/3 of the Guadua related research proposals since 2001 were approved and funded by COLCIENCIAS. Five out of nine approved research projects receive funds out of the programme 'Basic investigation in development of science and technology' (in this case four micro-level propagation projects and one related to silviculture). The remaining four projects are funded by the 'Innovation and entrepreneurial development in the agricultural sector' programme (three projects related to technologies and research for use in social housing projects and one related to the kiln/solar drying techniques of Guadua strips)[606]. All of the three neglected research proposals concentrated on the industrialisation of Guadua products. The reason for rejecting these proposal was lacking economic feasibility.

[606] In *annex 9* the total number of all approved COLCIENCIAS projects from 1993 to 2003 are listed.

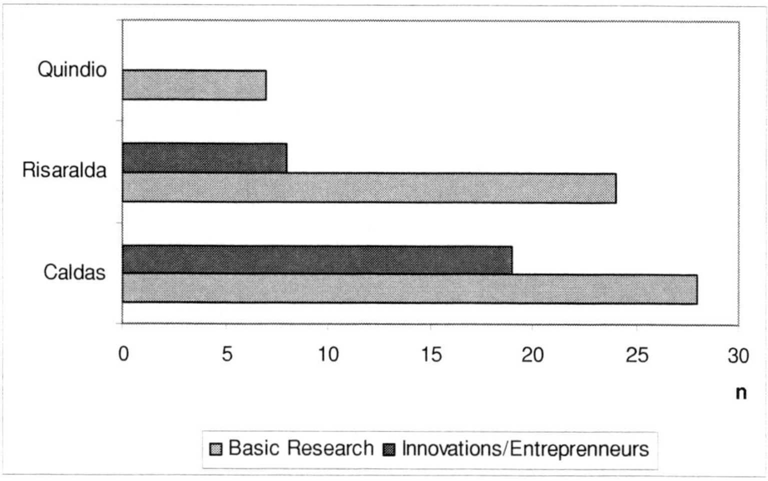

Graphic 9. Total number of COLCIENCIAS funded research project subdivided by research focus in the three departments of the study region (1993-2003)

In the study region the number of COLCIENCIAS funded research significantly increased since 2000. Before 2002 and in 2003 there was almost no research funded by the 'Innovation and entrepreneurial development' programme. The department of Caldas accounts for the highest number of COLCIENCIAS funded projects, whereas the department of Quindio does not receive any funds for projects of the "Innovations and entrepreneurial development" programme. Caldas is the oldest of the three departments and has the highest density of universities and research institutions. Quindio on the other hand is the most rural department. However, it accounts for the largest area of Guadua resources in the study region. The detailed figures of these developments are shown in *graphic 9*[607].

One important organisation for the industrialisation processes in Colombia is the national standardisation organisation ICONTEC[608]. It is a key partner of COLCIENCIAS and its members in order to coordinate NIS related activities and to achieve nationally harmonised and internationally standardised R&D results. ICONTEC started an initiative for the standardisation of Guadua applications and products in 2002. In 2003, an expert group was formed to develop norms for Guadua product groups. The group split up into subgroups consisting of experts for different aspects of Guadua production and processing. These expert groups are made up of entrepreneurs and researchers as well as of representatives of governmental organisations. Right now there are 5 to 6 of these groups working on proposals for the standardisation of products and production processes, for example the drying and

607 For a detailed listing of annually funded projects per department per programme from 1993 to 2003 see *annex 9*.

608 Spanish: *ICONTEC=Instituto Colombiano de Normas Técnicas y Certificación*

preservation of raw material or the production of laths and laminated bamboo. Two of these proposals are currently openly discussed[609]. The proposals will serve as official ICONTEC technical guidelines for the production and manufacturing process. They are not meant to become obligatory production norms. The technical guidelines will inform the producers about the best practice of a production process. Additionally, the guidelines will ensure international competitiveness, since they consider international standards of the different product groups.

6.4.2 Representatives of Guadua related Programmes and Initiatives

The national programme '*Manejo Sostenible de Bosques en Colombia*[610]' (MSBC) was decisively responsible for the development and adoption of the common utilisation regulation of the departments in 2001[611]. However, in 2003, the objectives and the scope of the activities of the national forest project MSBC have been redefined. Amongst others the 'new' MSBC strengthens the role of local (municipal) user group organisations, resulting in the formation of four user group associations at the municipal level[612]. The results of the newly defined role of the MSBC are described in *chapter 7*, which deals with the direct promotion of micro-level actors in the Guadua sector. The most important institutional changes of the MSBC Guadua programme were:

- Combination of the Ministry of Environment with the Ministry of National Housing and Territorial Development
- Decreasing importance of the environmental aspects in the national policy
- Reduction of CAR activities as defined in the Colombian environmental law of 1993 (administration and control)
- Stronger integration of municipalities in environmental projects and land use concepts (i.e. in Guadua related MSBC projects)
- Reduction of staff responsible for natural resources due to decreasing national environmental budget
- Utilisation, processing and marketing of forest products is within the responsibility of the Ministry of Agriculture and Rural Development" (Law 101 of 1993 and Law 811 of 2003)

[609] ICONTEC Proposal No. DE 549/03: Harvesting and post-harvesting guidelines for Guadua and ICONTEC Proposal No. DE 225/04: Drying and preservation of Guadua raw material.
[610] Transl.: Sustainable Management of Colombian Forests
[611] This regulation is the '*Norma Unificada en Guadua*' as it is described in *chapter 4.4.1*
[612] As already happened in the municipalities of Quebrada Negra and Marsella, see also *chapter 7.3.1*.

The second actor to be described in this chapter is the National Guadua Council (*'Consejo Nacional Guadua'*), which was formed in 2003. The council was established on the basis of the Colombian Law 811 (2003) which reformed and redefined the role of the productive chains in the agricultural sector[613]. The council consists of departmental councils, which in turn form a national roof council. Basically, the core council consists of private actors and entrepreneurs, which have a special interest in the use and commercialisation of the Guadua resource. This organisation is still in its initial phase and is closely cooperating with organisations such as the IICA (*'Instituto Interamericano de Cooperacion en Agricultura'*), NGOs, chambers of commerce, SENA, CARs, MSBC, etc.

The council consists of Guadua related entrepreneurs from the departments. They were elected as representatives of the respective department during the first meeting of the council, facilitated by the CARs and other governmental and non-governmental organisations. The entrepreneurs, who form the Guadua council are not necessarily the biggest or the most experienced or successful ones. Rather, they are owners of SMEs and come from various backgrounds, e.g. the wood processing sector, mechanical engineering, and others. The council is made up of approximately 20 entrepreneurs and a national secretary, who is in close contact with the Ministry of Agriculture. The meso-level actors offer technical and institutional support to the council. It is still discussed, whether these meso-level actors shall be entitled to vote in the Guadua council.

6.4.3 Representatives of the Forestry and Wood-processing Sector

The forestry and wood-processing sector of Colombia was investigated in order to evaluate its strategic positioning with respect to the Guadua sector. From the forestry sector the following meso-level representatives could be interviewed:

- Ministry of Environment *(Ministerio de Medio Ambiente, Vivienda y Desarrollo territorial)*; Department for the development of the national forest plan
- CONIF (Corporación nacional de investigación y fomento forestal)

Interviews were also sought with private associations and entrepreneurs of the wood and wood-processing sector. This aim could not be achieved due to the currently ongoing institutional and structural changes. Contacted potential interview partners articulated disinterest in the Guadua sector. However, the following aspects of the forestry and wood-processing sectors were examined drawing upon information provided from written sources and the CONIF and the Ministry of Environment:

[613] See *chapter 4.3* on productive chains in Latin American economies.

- Relations between the Guadua and wood sector
- Competing or complementing situation/perceptions
- Common/synergetic perspectives
- Potential fields of co-operation: technology, marketing, research, organisation

The Guadua sector in the coffee growing region must be seen within the wider national context of the wood and wood processing sectors of Colombia. Guadua as possible substitute and supplement for traditional wood consumers should at least draw the attention of the actors of these sectors. The wood and wood-processing sector in Colombia is not organised into productive chains as compared to other agricultural sectors.

In Colombia, as in many other Latin American countries, the productive chain approach is applied in order to consolidate the various producing sectors and improve their national and international competitiveness (Colombian Law 101 of 1993). This approach implies that the private actors of the respective sectors are responsible for the general entrepreneurial and strategic decisions. At the same time the state provides the necessary institutional framework, incentives and subsidies, and co-finances research.

The first productive chain approach in the Colombian wood processing and forestry sectors was introduced in 1996 and included the pulp and paper related enterprises, including the providers of the raw materials. During the field work for this study in 2002, the Colombian pulp and paper (P&P) industry was already taken into account (one interview was done with Colombia's largest pulp and paper producing company). Due to technological decisions made in the past (dissolution processes and technological manufacturing equipment), bamboo was not considered to be appropriate for pulp and paper production. Furthermore, it was mentioned that the P&P sector's timber resources, as well as the resources of the providers of the sector are sufficient and cheap enough. Therefore, there is no need to change the utilised resources.

The second productive chain approach was initiated in 1998 and renewed in the year 2000. The aim was to include timber producing and timber processing industries (timber, furniture and other wood products). The focus of these consolidation approaches was on timber production based on wood extracted from plantations and secondary forests, predominantly in four departments of Colombia. The Colombian Ministry of Agriculture classified these departments as 'Core Areas of Forestry Related Activities' (*Nucleos Forestales*). Around the forestry and wood processing industries in these *Nucleos*, a structure of facilitating and supporting organisations was established. However, the national productive chain of the forest and wood sector never reached advanced organisational and institutional consolidation.

In the year 2003 the Law 811 was introduced, which reformed the productive chain approach of the past and resulted in a complete re-organisation of the institutional structures of the forestry and wood processing sectors. In November 2003 a national committee was formed and drafts for a common productive chain strategy are in preparation.

In order to get an idea of the perceptions of wood and wood-processing actors and their attitudes towards the Colombian bamboo sector, organisations involved in the consolidation process of the Colombian forestry and wood processing sectors were interviewed. These organisations were: the department for planning of the national forestry plan and the national wood and wood-processing research institution (CONIF). The most important statements made by these entities were:

- Neither the Guadua-actors nor the actors from the timber and wood-processing industry feel the necessity to combine forces and resources in an association.
- However, it was recommended, that both productive chains sign an agreement on co-operation and communication.
- During the whole reform process of the timber and wood-processing sector the Guadua-topic has never been mentioned.
- Explanations of the wood-sector for neglecting the bamboo sector possibly are: Small local/regional focus of the Guadua topic and sector, small resource base, the artisan level of production of the bamboo sector.
- Further explanation of the situation is, that both sectors, the wood and bamboo sector have low levels of organisation.
- The wood sector does not perceive the bamboo sector as competition, rather as potential provider of raw material for further processing (namely flooring).
- The wood sector itself is highly heterogeneous and interests are widely spread, i.e. large-scale reforestation companies, the pulp and paper industry and the furniture producers.
- The bamboo sector, which is much more dependent on small and medium-sized enterprises, would feel inferior in potential institutional co-operations or merges. Furthermore, resource distribution would favour the wood industries and research institutions.

7. Research Results: Interactions between the Meso and the Micro-Level

The theoretical discussion of the factors, which underlie increased competitiveness of an economic sector in *chapter 3* concludes that the patterns of interactions between the actors of the meso and the micro-level play a crucial role in the upgrading processes of economic sectors. This chapter highlights the interactions between the two levels.

As presented in *chapter 6.2.3*, the basic aim of the meso-level actors is the promotion of industrialisation processes in the Guadua sector in order to increase the sector's contribution to the study region's social welfare and GDP. To achieve this aim, the meso-level actors in the sector adopt promotion activities, which can be subdivided into the three categories R&D, E&E and I&I[614]. Each of these three categories implies that the meso-level activities are at least partly linked to the direct interaction with the micro-level actors

Table 35 shows the interactions between the meso and the micro-level for the three categories. E&E activities are the most common means of micro and meso-level interactions. R&D and I&I activities are usually communicated and delivered to the micro-level through the use of E&E related activities. The indirect communication through R&D and I&I are preferred over the direct integration of micro-level actors into ongoing activities. In the following sections of this chapter, the three fields of activities and the respective meso and micro-level interactions are described, using examples as they were encountered during the field work in the study region in the years 2002 and 2004. The following research questions served as a guideline for the analysis:

- How do meso-level actors integrate micro-level actors into their activities?
- How do the meso-level actors try to enhance the innovative capacity of the micro-level actors?
- Is it possible to draw conclusions from the interaction behaviour among the meso-level actors for the way they interact with the micro-level actors?

[614] R&D= Research & Development; E&E= Education, Dissemination & Expansion; I&I= Creation of Institutions & Promotion of Interaction (see *chapter 6.3*).

Table 35. Interactions between the meso-level and the micro-level in R&D, E&E and I&I

	Meso-level					
	R&D Activities		E&E Activities		I&I Activities	
	Direct	Indirect	Free	Fees/Costs	Participation	Top-down
Interaction	- Integration of micro-level actors into R&D, e.g. through on-farm research or product development in co-operation with private entrepreneurs	- R&D mainly among meso-level actors - Dissemination of R&D results and experiences through E&E activities. E.g. through E&E modules in R&D projects - Through the application of I&I activities, such as the provision of credit schemes to enhance entrepreneurial R&D	- E&E-activities are conducted free of charge and open to everybody or to a target group	- The E&E activities generate income for the respective meso-level actor. The barriers to participate are usually inscription/enrol-ment or participation fees	- Integration of micro-level actors into institutions-building processes, e.g. the integration of private experts in the development of product norms or regulations - The facilitation and financing of micro-level organisations and projects	- I&I activities are elaborated only amongst meso-level actors as e.g. during the preparation and adoption processes of many laws - Dissemination and facilitation of I&I results and experiences through E&E activities
	Micro-level					

7.1 Research and Development (R&D-Activities)

In *chapter 6.3.3* it is shown that the R&D activities account for the highest number of intra-meso-level co-operations. Mainly the research institutions, the public-private actors, as well as the governmental actors are active in this field. Regarding interactions in the area of R&D the meso-level actors did not mention private enterprises. They rather mentioned the co-operation with independent experts, whom they usually employ for research activities. During the research in the years 2002 and 2004 in the study region, only few co-operations with micro-level enterprises and farmers could be identified.

Considering the Guadua sector problems the interviewed meso-level actors mentioned especially the inter-organisational communication and co-operation problems, as well as the deficient publication of R&D results and experiences[615]. On the other hand, the meso-level actors stated that micro-level actors lack the technical know-how and capacity to produce high value Guadua products and compete in the national and international markets (see *graphic 10* for the list of problems of the Guadua producing and manufacturing sector as perceived by the meso-level actors). The meso-level actors identified promising potentials of

[615] See *chapter 6.2.1*, *graphic 3*

Guadua and Guadua products in the domestic and the international markets, on which R&D activities should focus according to them. These products and potentials are listed in *graphic 11*. Basically, the meso-level actors pursue the strategy to conduct R&D activities with partners at their own level and disseminate the results through E&E activities to the micro-level later on. The micro-level is perceived as a receiver of such efforts rather than a potential partner for R&D activities. Exceptions are a few large-scale commercial bamboo producers, who maintain lose, informal contacts to research institutions and governmental actors (see *chapter 8.4*).

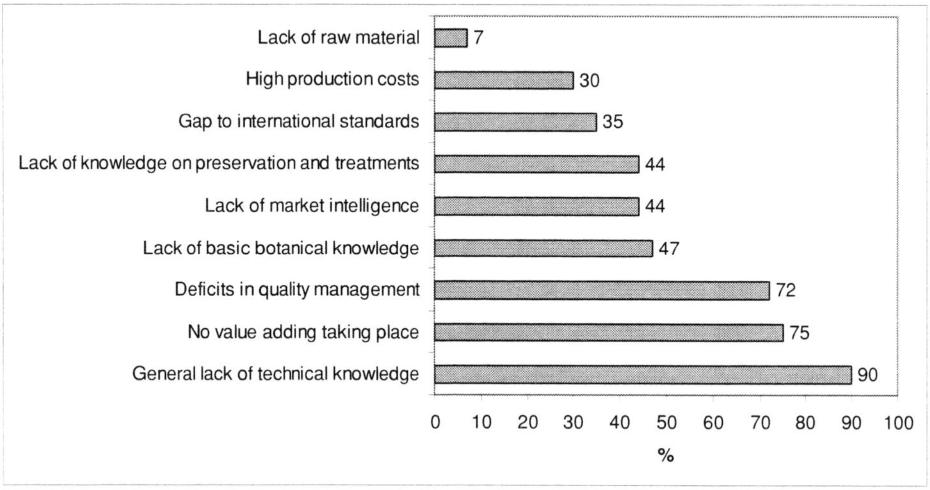

Graphic 10. Major problems of the Guadua producing and manufacturing actors according to meso-level actors (n=43)

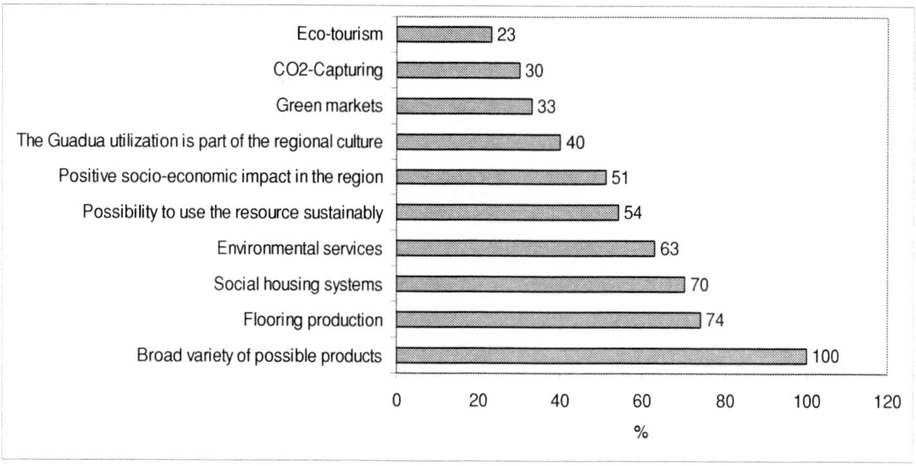

Graphic 11. Major potentials of Guadua and Guadua products according to meso-level actors (n=43)

In the *net-diagrams 1-6* and in *chapter 6.3.1*, the foci of the meso-level R&D activities are shown, indicating that product development and market research are the dominant activities, whereas basic research and process development do not receive as much attention. With respect to the integration of micro-level actors, on-farm research of silvicultural treatments and propagation trials are very common. In the product and process development activities, the integration of entrepreneurs has not yet reached a significant level. No formal co-operation agreement between meso-level actors and private enterprises was found. Nevertheless, some private entrepreneurs share their knowledge about products and processes at conferences and as instructors in educational courses organised by the meso-level actors. In rare cases these entrepreneurs provide their knowledge and sometimes even old machinery to micro-projects, such as communal enterprises for raw Guadua processing funded by one or several of the meso-level actors (see *chapter 7.3.1*). Another link between the meso and micro-level is that individual researchers and technicians from meso-level institutions start their own Guadua related businesses; a process, which is important for the spreading of R&D results within the sector. However, this phenomenon could not be investigated in depth, since such enterprises are still scarce and not yet established.

As mentioned before, the meso-level engages in indirect R&D promotion at the micro-level rather than integrating micro-level actors directly into R&D activities. However, entrepreneurs may apply for R&D supporting funds. These funds are provided by the National Innovation System (NIS) in form of credits and financing schemes to enhance entrepreneurial R&D activities. COLCIENCIAS funds private entrepreneurs only if they co-operate with meso-level actors, such as universities, NGOs, governmental actors, etc. As *chapter 6.4.1* reveals, the number of COLCIENCIAS funded Guadua related projects is still small. However, the figure is increasing, particularly in the field of projects related to entrepreneurial development and innovation.

The analysis of the meso-level actors promoting small and medium-size enterprises revealed the disadvantage of Guadua related SMEs with respect to acquiring institutional support and funding. The ANDI[616], which is a private association of medium and large-scale entrepreneurs, was never approached by Guadua enterprises, nor was the ANDI approached by the wood-processing industries in the region, although ANDI states, that any industrial enterprise is welcome to apply for technical and institutional support.

The INCUBAR or INCUBADORA[617] networks provide more feasible and available support for Guadua related enterprises. These types of organisation officially belong to the Colombian National Innovative System (NIS). Basically, the INCUBADORAS provide entrepreneurial aid for recently founded enterprises. For enterprises, which have been accepted into the programme, the SENA or the Chambers of Commerce organise extension activities. They provide support in the development of business plans and access to capital and credit schemes. The INCUBADORAS facilitate institutional and entrepreneurial coordination, communication and consolidation activities. The most important support the INCUBADORAS offer to the enterprises is the establishment of relations to private investors. Since the economic experts of the INCUBADORA must approve of the business idea and because the INCUBADORAS provide institutional, technical and structural support to the new enterprises, the risk for the private investor is reduced significantly. The major advantage of this financing scheme compared to bank and state credits are 1) the involvement of private market participants with their experience and expertise and 2) a more flexible design of the credit conditions.

Both interviewed INCUBADORA networks said, that they plan to support Guadua related enterprises in the near future. This support will be of a structural nature and will be implemented in co-operation with other organisations in the region. They also mentioned, that in the past, several business ideas about Guadua processing had been proposed. However, these enterprises did not enter the promotion programme for two major reasons: lack of economic feasibility (possible outlets and production costs) and the artisan character of the enterprises. To understand the decisions of the INCUBADORAS a list of the most important selection criteria are compiled below:

- Profit orientation
- Free participation in the market
- Innovative business idea (the term innovative was related to the product or service offered by the enterprise)

[616] Spanish: ANDI=*Asociación Nacional de Empresarios de Colombia*. In the Colombian national entrepreneurs association about 650 medium and large-scale enterprises from all economic and industrial sectors are organised. For details see www.**andi**.org.co.

[617] Spanish: INCUBAR/INCUBADORA=Incubate/Incubator. For a detailed description of their work and their distribution in Colombia see www.incubarcolombia.org.co.

- Feasible business plan and market analysis
- Regular monitoring and reporting

According to the INCUBADORAS, the most preferred and accepted business ideas in the last year were: e-commerce and web-design; machinery; textiles and some extraordinary ideas from the agricultural sector. According to INCUBADORAS, the Guadua sector inheres a number of potentials, but the enterprises of the sector lack the innovative capacity and the economic perspectives of other sectors and business ideas.

In the innovation and product development processes, the investment capital for new machinery, to pay employees, to cover costs of product development, to purchase materials, etc. are seen as major obstacles especially for SMEs. In the study region banks and NGOs were interviewed regarding their offer of credit schemes for small and medium-sized enterprises in the region. *Table 36* shows that at least according to the knowledge of the interview partners, no Guadua related enterprise received credits from the group of interviewed banks. Moreover, the interviewed people stated, that they do not remember one case of a Guadua related enterprise, which applied for a SME or FINAGRO[618] credit (the latter is the official governmental credit line especially for agricultural and agro-industrial SMEs). From the two interviewed NGOs, one is actively involved in the promotion of the Guadua sector, which is implemented through a multi-actor co-financed project (see *chapter 7.3.1*).

The interviewed representatives of the credit organisations were asked for the most successful business proposals from small and medium sized enterprises regarding acquirement of credits. Banks and NGOs commonly named enterprises related to internet commerce and web page design. Second in this ranking were enterprises from the textile sector and the third place was held by enterprises from the metal processing sector. Furniture production was also mentioned as a promising business idea. However, all interviewees agreed that the wood and wood-processing sector in the region is generally not seen as a promising business option. Moreover, the wood producing enterprises (reforestation companies) have a rather negative image regarding compliance with credit conditions[619]. Reforestation is seen as an activity, which should be subsidised or at least pre-financed by the state.

---

[618] Spanish: FINAGRO=*Fondo para la Financiación del Sector agropecuario*. The FINAGRO offers two-year credits, which cover up to 100% of the costs related to production of agro-industrial products. The beneficiaries are small producers, which are defined according to the assessed value of their property (in 2001 Col. $ 33,975,000, which is around € 17,300). In the year 2003 the credit scheme released Col. $ 514,000,000 (€ 150,600). For more details see www.bancoagrario.gov.co and www.finagro.gov.co.

[619] The Colombian author SALOMON KALMANOVITZ sees the "*Cultura del no pago*" (the culture of not paying) as a general problem of Colombia's society and as a major obstacle to economic development. This refers to the non-compliance of all types of entrepreneurs and individuals with the fiscal and tax system and the unreliability regarding the compliance with the conditions of private and state credits, as well as paying back debts (see KALMANOVITZ, 2001a: 100ff.).

Table 36. Characteristics of credit schemes offered by banks and NGOs relevant to the Guadua sector

Interviewed Actor	Actor offers own SME-credit	Actor offers state SME-credit	Actor offers FINAGRO credit	Actor released credits for Guadua enterprises*
Bank 1	Yes	No	No	No
Bank 2	No	(IFI 2003)[620]	Yes	No
Bank 3	Yes	(IFI 2003)	No	No
NGO 1	Yes	No	No	Yes
NGO 2	Yes	No	No	No

*The bank/NGO granted credits to Guadua related enterprises from one of the three named credit schemes

The official state credit schemes are covered and administered by the national Colombian BanColdEx[621] Bank, which distributes the money and which determines the requirements, which the credit receivers must fulfil. The final approbation of the credit, however, lies within the responsibility of the individual banks. The risk is shared between the two banks and the debtor according to the changing conditions. These official state SME credit schemes (such as the IFI), have one important advantage, which closes an important gap for SMEs to have access to investment capital. In the day-to-day credit lending process, 'normal' credits require collaterals of the debtor. Usually, SMEs do not account for a sufficient stock of capital or property to comply with this lending condition. The Colombian state offers a national fund to provide collateral guarantees for indigent debtors; the '*Fondo Nacional de Garantias*'. But, access to this fund is bound to the approbation of the credit by the respective bank.

According to the interviewees the major obstacles for Guadua related enterprises to receive bank or public credits are (in order of importance): The informality of the enterprises; the size of the enterprises (micro enterprises); entrepreneurs of the sector do not know of their credit possibilities; too much bureaucracy and high transaction costs for these types of enterprises; no economic feasibility of the enterprises; lacking management and accounting skills; negative side effects for the debtors (e.g. registration in chambers of commerce, tax duty, etc.)[622].

---

620 Spanish: IFI=Instituto Fomento Industrial (Institute to enhance industrial development). For more details see www.ifi.gov.co. The IFI especially supports SMEs on their way from artisan production to more industrialised processing techniques.

621 Spanish: BanColdEx=Banco Colombiano de Exportación. For more details see www.bancoldex.com.

622 CLARKE ET AL. (2002) see in the increasing involvement of international banks in Latin America a reason for increased loans to SMEs. In Colombia, where especially small foreign banks are active, the situation is different since these banks tend to lend less to SMEs. For this research the shareholder details of the investigated banks could not be revealed. Due to this and the small number of investigated banks CLARKE's findings could not be verified or denied with respect to the Guadua related enterprises.

7.2 Extension, Dissemination of Knowledge and Education (E&E Activities)

Over the past five years there has been an extraordinary increase of interest in the Guadua sector and with it an increase in Guadua related activities and actors. One field of activity, which is especially affected by this development are the E&E activities. Compared to the year 1997, amongst the interviewed meso-level actors the number of those who are actively involved in Guadua related extension and professional education has more than doubled. However, the co-operation rate between the meso-level actors in the area of E&E activities is rather low.

In the Guadua sector the most important E&E activity is the technical extension. It is usually free of charge and fess and has three foci 1) Propagation and silviculture, 2) Constructions and buildings, and 3) Handicraft and furniture made of Guadua. Most recently, new topics, such as wood preservation and wood drying techniques or the fabrication of Guadua laths have also been included.

The professional education about Guadua focuses on subjects, such as architecture, civil engineering, environmental administration and silviculture. The universities in the study region offer these courses to their students and staff. Additionally, there are courses and seminars for professionals and other interested people, held on weekends, in the evenings and during the official vacation time. A fee has to be paid for these courses and participants usually receive certificates at the end. Apart from the tuition fees of the regular students, this source of income contributes substantially to the universities` annual budget. According to the interviewed university representatives popular topics attract larger numbers of participants. The universities contract well paid experts, such as the experts described in *chapter 6.3.5* to improve the course quality and its reputation. In the sector of professional education the situation can be described as competitive between the education entities, which possibly explains the low rate of co-operation in the field of E&E.

The entrepreneurial extension includes of business formation, business management, taxation, employment and other legal aspects of business administration with focus on the specific requirements of the Guadua based enterprises (e.g. in the taxation system of natural resources and its manufactured products of the export of such products). Many extension and education activities lack this important component, which is confirmed by the high number of informal Guadua based enterprises without basic patterns of business management (compare *chapter 8*).

Most meso-level actors are involved in PR and dissemination activities. Nevertheless, in the problem statements presented in *chapter 6.2.1* they complain about the general lack of publications and information on Guadua related activities and research results. According to the experiences made during the fieldwork, this information and knowledge is available. However, it is scattered all over the region and the country. It is difficult to get access to the information since a number of administrative and institutional obstacles, such as formal letters of request, meetings with the responsible person to justify the insight, etc. must be overcome. These obstacles have negative impacts on the availability of information, especially to micro-level actors.

According to the interviewed meso-level actors several hundred people were educated over the past five years. The most popular extension and education courses and seminars are currently offered by the SENA and the universities and deal with Guadua constructions. Fees must be paid and professionals from various backgrounds do these courses. SME entrepreneurs are hardly found amongst them. The NGOs and governmental actors mainly offer courses on silviculture and technical development to promote the production of furniture and handicraft. The beneficiaries of these courses are farmers and unskilled newcomers to the Guadua sector. Already established formal and informal enterprises mainly benefit from E&E activities by chambers of commerce and trade and other public-private actors, which offer a mix of entrepreneurial extension and technical extension.

7.3 Institutions-building and Initiation of Projects (I&I Activities)

In the following, the I&I activities at the meso-level are analysed. Focus is their impact on the micro-level and the integration of the micro-level actors into the meso-level's I&I related activities. I&I activities can be divided up into 1) Institutions building and 2) the promotion of the sector's organisation and interactions (co-operations) between the micro-level actors, namely through micro-level projects.

The meso-level actors initiated the creation of a number of institutions. The most relevant of these processes are: 1) The MSBC, which established a regulation for the common use of Guadua resources in five departments, replacing the differing previous regulations. 2) The ICONTEC, which aims at the establishing of product and production norms and quality guidelines for Guadua products and constructions. 3) The National Guadua Council, which is spearheading the organisation and consolidation of the Guadua sector and its recognition as an official Colombian productive chain in the agricultural sector. The meso-level actors were leading figures in the initiation of the three processes. However, micro-level actors were assigned central positions, since the processes demand technical and entrepreneurial expertise and consultation. In *table 37*, the three processes and the role of the integrated micro-level actors are described.

Besides the on-going institutions-building processes, there are already institutions in place, governing and influencing the use and manufacturing of the Guadua resources in the study region. In *table 38*, the dominant institutions influencing the Guadua sector are presented. The most important ones are printed in bold letters. The three institutions building processes presented in *table 37* aim at the modification and amendment of the major institutions to adopt more to the Guadua sector's conditions and development goals. This includes: 1) The recognition of the Guadua sector in the national framework institutions manifested through the establishment of a Guadua productive chain according to the agricultural law (National Guadua Council); and 2) the codification and improvement of traditional/informal knowledge on resource management and application of the raw material (ICONTEC and MSBC).

Table 37. Institutions building processes and the integration of micro-level actors

Process	Objectives	Leading Meso-level Actors	Integrated Micro-level Actors	Role of Micro-level Actors
MSBC (Pilot region "*Eje Cafetero*")	- Legal regulation of the use of Guadua resources - Adoption and application of the environmental law to Guadua related aspects	- CARs	- Guadueros - Large-scale farmers	- Knowledge and experience - On farm research
ICONTEC	- Quality norms and production guidelines - Amplification of building and construction codes and product quality norms considering the characteristics of Guadua materials	- ICONTEC - SENA - Universities - SCB - CARs	- Civil engineers and architects - Experienced micro-level entrepreneurs (experts)	- Development of proposals for standards and guidelines to be considered during the ICONTEC decision-making process
National Guadua Council	- Establishing a Guadua productive chain according to the Colombian agricultural law	- CARs - Universities - Agricultural research institutions - Agricultural Producer Organisations	- Entrepreneurs and experts from agriculture, silviculture, manufacturing, etc.	- Round tables with meso-level actors to develop common proposals on how to organise the sector and where to put the strategic emphasis

Table 38. Institutions influencing the Guadua sector

Overall Framework	Agricultural Sector	Forestry Sector	Guadua sector	Promotion of SMEs
Environmental law; National and international commercialisation of natural products; General building and construction codes; Export promotion; Fiscal system; Labour legislation; Business legislation;	**Agricultural law on 'Productive Chains';** Management of forest plantations	**Forest law; Management of natural forests;** Forestry sector agreements; Forestry productive chain approaches	**Legal regulation of the management and use of Guadua resources; Extension and education of Guadua producers and manufacturers; Common (codified) knowledge on constructions and processing techniques of Guadua;** Traditional management and use of Guadua stands; Basic construction codes for Guadua building elements	Promotion and formalisation efforts for SMEs; Micro-credit schemes of NGOs, private banks and the state; Extension and education for entrepreneurs

7.3.1 Analysis of Guadua based Projects

A second important strand of I&I related activities is the promotion of the micro-level's organisation and co-operation capacity. This is usually achieved through the facilitation and financing of micro-level projects, either within a general programme of a broader context or as part of an individual Guadua related micro-project. The following sections present an overview of the results derived from the analysis of four Guadua related projects in the study region. Two of the projects are conducted within a national programme, which provides funds and technical expertise. It is the FOMIPYME programme[623], which focuses on the promotion and creation of micro-enterprises in any economic sector. The other two projects both emphasise the establishment of Guadua related community-based enterprises. These two projects receive technical and entrepreneurial support from the MSBC programme. The local municipalities make funds and infrastructure available.

The four projects were visited in the year 2004. To get an overview of the capacity and the progress of the projects, open-end interviews were conducted with leading project members during field trips to the project sites. Since the earliest project only started in late 2002, the projects` 'success' or 'failure' could not be evaluated. In *tables 39a* and *39b*, the basic

[623] FOMIPYME=Fondo Colombiano para la Modernización y Desarrollo tecnológico de las micro, pequeñas y medias Empresas. For details see http://www.bancodebogota.com.co/decreto.pdf.

features of the four projects are listed and compared. The presented criteria were chosen in order to present an overview of the projects. Emphasis was put on the type of micro-level project members, the state of products and production capacity and the institutional frameworks of the projects (i.e the support by meso-level actors).

Table 39. Micro-level Guadua projects in comparison

Project type	Micro-level members	Products	Production technology	Institutional embedding	Long term objective
1) MSBC	- Landless labourers of the coffee region, inexperienced in Guadua management and processing	- Standard Guadua raw material - Guadua laths	- Sorted out machinery, constructed and tested by experts	MSBC, municipality and universities provide: - Technical expertise - Infrastructure - Access to credits	- Self sufficient community based enterprise with focus on raw material for construction and pre-fabricated products
2) MSBC	- Labourers and small scale farmers of the coffee region, mainly inexperienced in Guadua management and processing	- Standard construction material	- Not yet purchased	MSBC, municipality and universities provide: - Technical expertise - Infrastructure - Access to credits	- Self-sufficient community based enterprise with focus on high quality construction material

Table 40. Micro-level Guadua projects in comparison

Project type	Micro-level members	Product	Production technology	Institutional embedding	Long term objective
4) FOMIPYME	- Large group of raw material producers provide material for a project owned factory producing laminated bamboo - Entrepreneurs experienced in wood and Guadua processing;	- From pre-fabricated products (laths) to high-quality end products (Laminated bamboo products)	- To be especially developed for this project	- Universities and public-private actors, such as chambers of commerce and industrial design laboratories provide: - Technical expertise - R&D facilities - Credit access	- Public-private partnership enterprises made up of a society of Guadua producers and processors, and state shareholders
3) FOMIPYME	- Inexperienced and experienced Guadua handicraft and furniture producers	- Modern Guadua products designed for the interior use in private households, offices, etc.	- Individual workshops of the members - Artisan character with high trial and error ratio	NGOs, Chambers of commerce and the handicraft design laboratory provide: - Technical expertise - Access to credits	- Improvement of the individual entrepreneur's competitiveness and product quality

Comparing the two types of project designs and the respective institutional support, it becomes clear, that both, MSBC and FOMIPYME, project designs aim at education of a large number of actors, preferably actors, who belong to a certain target group (for example farm workers or micro-enterprises). The major differences between these two promotion programmes are: 1) The MSBC projects emphasises the role of the natural resources for the creation of employment and regional development; 2) The FOMIPYME programme supports SMEs of all economic sectors. Thus, the Guadua related projects compete with other types of projects for this support.

The focus of MSBC projects is put on the provision and processing of raw materials for the construction sector. The target group consists of landless farm workers, who manage and harvest the Guadua of the farms in their region. They add value to the raw material through well-known treatment processes and they offer products, which are easy to produce and process in great quantities. The project members usually do not have any education regarding Guadua management and processing or the management of an enterprise. The MSBC assists with technical and entrepreneurial education and facilitates access to credits and funds. These activities are supported by the municipalities, which provide infrastructure and funds. Aim of the MSBC projects is to support the initiating phase and to create self-sufficient co-operative associations in the long run.

The FOMIPYME projects aim at the improvement of already existing utilisation, enterprise and commercialisation structures. The entrepreneurs receive economic education and technical support in order to improve their product and production quality, as well as to integrate them into the formal economic sector. FOMIPYME represents a contract between a number of meso-level organisations, which commit themselves to provide support and assistance to a selected group of micro-level actors according to their organisations` expertise. The FOMIPYME fund pays for 50% of the arising costs, whereas the remaining 50% are covered by the members themselves. Aim of each FOMIPYME project is to educate entrepreneurs in order to make them more competitive integrate them into the formal economy. In contrast to the MSBC approach it supports individuals rather than co-operatives.

8. Research Results: Interactions and Innovations at the Micro-level

This chapter presents the results of the investigations conducted amongst micro-level enterprises of the Colombian bamboo sector. Special emphasis is put on the innovations of these enterprises and how the innovations are affected by the meso-level promotion activities. In order to do so the sources of and reasons for innovations of the enterprises are analysed in-depth. In the course of the research 186 micro-level actors were interviewed (see *table 40*). Of these, the innovative behaviour and the patterns of interaction of 48 handicraft producers, 12 furniture producers, and 6 special providers of Guadua products were analysed. The following sections present the results of the analysis. The research results related to the other micro-level groups of actors (the farmers[624], intermediaries, the retailers and the constructors) were published by HELD AND MANZANO (2004) as a working paper (No. 51) of the International Network for Bamboo and Rattan (INBAR).

Table 41.Investigated groups of actors at the micro-level

Micro-level group of actors	Number of interviewed enterprises/actors
Coffee-Farmers*	59
Commercial bamboo producers	6
Intermediaries (Guadueros)*	2
Distributors/Retailers*	30
Constructors*	18
Furniture Producers	12
Handicraft Producers	48

* For an analysis of these groups of actors see HELD AND MANZANO (2004)

The aim of the micro-level analysis of the Colombian bamboo sector was to answer the following set of research questions:

- How do SMEs in the Colombian Guadua sector innovate?
- Which influences have the interactions between SMEs on the innovative behaviourof the entrepreneurs?
- Which influences have the meso-level's promotion activities on the micro-level's innovations?

[624] The interviews and the basic data analysis were done by the Colombian NGO CIPAV in the year 2002.

8.1 Informality and Innovations in Guadua based small and medium size Enterprises (SMEs)

The Guadua sector in the Colombian coffee growing region was visited two times, in 2002 and 2004. HELD AND MANZANO published the basic research results of the 2002 investigations in 2004. For the analysis of the Colombian bamboo sector they adopted the production-to-consumption-system analysis (PCS) approach, developed for the International Network for Bamboo and Rattan (INBAR) in 1994 by BELCHER. This approach is meant to serve as a framework for socio-economic case studies in the bamboo and rattan sectors of the world and involves: [625]

- Definition of the boundaries of the production-to-consumption system;
- Assessment of state of the art of technology and production in the sector;
- Mapping of the various links between the actors and within the system;
- Definition and assessment of policies and the institutional environment, that guide and constrain the system;
- Assessment of constraints to knowledge, resources, and incentives;
- Definition of the specific changes that must be introduced;
- Definition of the policies and institutional interventions needed to achieve the changes.

The study of HELD AND MANZANO concludes that the Guadua sector in the study region is characterised by high entry and exit dynamics amongst the micro-level enterprises, the artisan character of the enterprises and the informality of the vast majority of the sector's enterprises. The sector lacks coherent structures and organisation. *Tables 41* and *42* deal with the enterprise structures in the furniture and the handicraft bamboo sub-sectors.

---

[625] BELCHER (1998)

Table 42. Basic structural data of the Guadua handicraft sub-sector 2002[626]

N = 48	Annual sales volume (in Mio. Col. $)				No. of employees					Machinery			Years in the business		
	5-10	11-20	21-50	n/a	1	2	3-10	=>11	Self-employers only	Wood processing	Modified wood	Only manual	<=3 years	4-10 years	>10 years
n	4	3	4	37	7	8	6	2	25	34	4	10	25	14	9

Table 43. Basic structural information of the Guadua furniture sub-sector 2002[627]

N = 12	Annual sales volume (in Mio. Col. $)				No. of employees				Machinery		Years in the Business		
	10-20	21-50	=>51	n/a	1	2	3-10	=>11	Only Manual	Modified wood processing	<=3 years	4-10 years	>10 years
n	2	1	3	6	3	4	3	2	5	7	3	4	5

The enterprises in the bamboo sector do not differ from enterprises in other Colombian economic sectors[628]. 49 out of the 60 investigated handicraft and furniture producing enterprises are not registered in the chamber of commerce and trade. These informal enterprises usually employ family members on an irregular basis without the provision of social services and healthcare. For some workers these enterprises also offer employment and income. More importantly, the workers acquire skills and knowledge on production processes and techniques and sometimes on marketing strategies.

---

[626] The table is based on data as presented by HELD AND MANZANO (2004). See *page xvi* for exchange rates.

[627] The table is based on data as presented by HELD AND MANZANO (2004). See *page xvi* for exchange rates.

[628] For the classification of micro-enterprises and SMEs in Colombia see law no. 590 of 2000 (REPUBLICA DE COLOMBIA. LEY NO. 590 DE JULIO DE 2000)

According to FERRANTI ET AL., in Latin American countries the major reasons for starting an informal enterprise are the lack of jobs for unskilled labour, a low level of wages in the industries and the independence from a superior[629]. These framework conditions can also be detected in the study region, where the coffee crisis of the 1990s seriously affected the labour market and the regional growth rates. The low level of minimum wages in Colombia has a negative impact on the formal labour market and enhances the formation of informal employment and self-employment[630]. In the Colombian rural areas, the income through informal employment and self-employment has increased with decreasing growth rates and wages[631]. But also in the larger cities, the informal sector is an important economic force, for example, in the early 1990s, the number of micro-enterprises[632] accounted for almost 90% of the total number of enterprises and for almost 40% of the employed labour in the city of Bogota[633]. The share of informal enterprises is not exactly known. Figures of the informal sector in Colombia let assume that more than 90% of the enterprises are not officially registered[634].

In Colombia more than one million informal enterprises exist, which create more than 60% of the national Colombian employment[635]. Since they do not contribute to the fiscal tax volume of the state[636] and by-pass the social welfare system, the aim of the Colombian national economic strategy is to incorporate these enterprises into the formal economy. For this purpose law 590 of 2000 was introduced[637]. This law offers the possibility to the informal enterprises to register without being immediately included in the tax system. Other important aspects of this law are a higher flexibility of the fiscal system regarding changing incomes and enterprise structures and the offer of extension activities[638]. In the Colombian bamboo sector, effects of the Colombian law no. 590 could not be observed.

The informality of enterprises does not only have negative effects on the national economy, such as the evasion of paying taxes, ignoring social services for employees or weak quality and production standards. The enterprises themselves are not entitled to state support, e.g. to subsidies or incentive and credit schemes, as they are not formally registered[639]. The

---

629 FERRANTI ET AL. (2001: 134ff.)

630 See BELL (1995: 20ff.) and BUSSOLO AND LAY (2003).

631 BUSSOLO AND LAY (2003: 25ff.)

632 According to the Colombian law no. 590 of 2002, a micro-enterprise is an enterprise with less than 10 employees.

633 URREA ET AL. (2000: 230)

634 GORDO (2002)

635 GORDO (2002: 10)

636 In Colombia the two most important sources of tax income of the state are the VAT (42% of the state's tax incomes in 2000) and the direct income tax (38%) (Internet database enquiry: REPUBLICA DE COLOMBIA. DIAN, 2004). Informal enterprises contribute to neither.

637 REPUBLICA DE COLOMBIA. LEY NO. 590 DE JULIO DE 2000

638 GORDO (2002: 14f.)

639 Most of the unregistered enterprises in Colombia (76%) lack the knowledge about the registration processes and the obligation to do so. Gordo (2002: 11). For a description of the broad range of sector and SME promotion activities in Latin America see GTZ (2001a) and LEVITSKY (1993: 48ff.).

economic situation of the informal SMEs and micro-enterprises is characterised by subsistence and survival oriented management patterns of survival cluster enterprises as described in *chapter 2.2*.

As the theoretical discussion revealed, the classical assumption is that enterprises` innovations are the result of R&D activities within their business structures. URREA ET AL. found that in 1996 only 25% of the Colombian enterprises were actively involved in R&D activities. The other 75% only acquired and modified well-known technology from the industrialised countries. Altogether, 75% of the Colombian enterprises invested in innovations in 1996[640].

For the Colombian Guadua sector, the entrepreneurial R&D activities were one of the unknown variables, which needed to be investigated. In innovation research the number of registered patents is an indicator for the innovation rate in an economic sector. An interrogation of the Colombian statistical patents database revealed only two patents for Guadua related products and none for process or technologies[641]. Thus, to investigate the entrepreneurial R&D activities and other sources of innovations in the Colombian Guadua sector, an indicator for innovative behaviour had to be defined. Such an indicator was product innovation. According to consulted experts the following product was introduced to the Guadua market only a few ears ago: an electrical lamp made of Guadua. This new product meets the requirements of an innovation, which is defined as: "*…a product, process or organisational change new to a firm or at least representing a significant improvement to it. This innovation has to be successfully established in the firm's market or has to be successfully in use in the firm*[642]".

Tables 41 and *42* illustrate that except for four enterprises, all of the 60 handicraft and furniture producing enterprises belong to the economic group of micro-enterprises as defined by the Colombian law. *Table 43* shows the correlations between the (in)formality of the enterprises and micro-economic features. Of the 12 furniture producers, seven are not registered, hence informal. Out of the 48 handicraft producers only six are registered for trade and commerce. Amongst the handicraft producers it is rather the new enterprises, which are registered (five of the six registered handicraft enterprises started their Guadua business less than three years ago). On the other hand, there is no clear correlation between the number of employees and formal registration. Four enterprises with one or two employees and two with ten or more employees are registered. The other enterprises with a moderate amount of employees are not registered.

---

640 URREA ET AL. (2000: 234ff.) based upon statistical data provided by COLCIENCIAS and the Departamento Nacional de Planeación (DNP)

641 Internet inquiry 2004. BANCO DE PATENTES. SUPERINDENDENCIA INDUSTRIA Y COMERCIO DE COLOMBIA. www.bancopatentes.gov.co.

642 OECD 1996

Among the furniture producers three of the four old enterprises (more than ten years old) are registered and two of the young ones (less than three years). There is a clear correlation between the size of the furniture enterprise and their registration. Except for one, all of the registered enterprises have more than three employees. Four of the five investigated enterprises in this category are registered. Regarding the annual sales volume and the legal status of the enterprises, the three largest furniture producers (more than 50 million Col. $/year) are registered. In contrast, no such correlation could be identified for the handicraft producers. Neither is there a correlation between the equipment of the handicraft enterprises and their registration. For the furniture producers again there is correlation between equipment and the registration at the chambers of commerce and trade (all the five registered enterprises belong to the group of seven enterprises using machinery).

Table 44. Correlations between enterprise (in)formality and micro-economic features of furniture and handicraft producers as assessed in 2002

SME Group	Formal/ Informal Enterprises	Registration/ Sales Volume	Registration/ Number of Employees	Registration/ Machinery Equipment	Registration/ Duration in Business
Handicraft (N=48)	6 formal/ 42 informal	No correlation	The two biggest employers are registered.	No correlation	Only enterprises younger than four years are registered
Furniture (N=12)	5 formal/ 7 informal	High sales volume correlated with registration	Larger employers are registered	Owners of machinery are registered	No correlation

After the analysis of the (in)formality of the investigated enterprises was concluded, the innovation behaviour of the investigated enterprises was focused on. Therefore, the basic micro-economic data was linked to indicators of innovation. In *table 44* the most reliable correlations are presented. The innovative firms were identified by means of their offered products. In case the handicraft producers named electrical lamps as most important product in terms of the sales volume, they would be included in the group of innovative enterprises. Eighteen out of 48 handicraft producers belonged to this group. For the furniture producers no such standard indicator was determined (*Graphic 12*). However, their innovative behaviour was determined by analysing the organisational, product and process changes, which were undertaken between the first survey in 2002 and the second in 2004 (*Tables 46 and 47*).

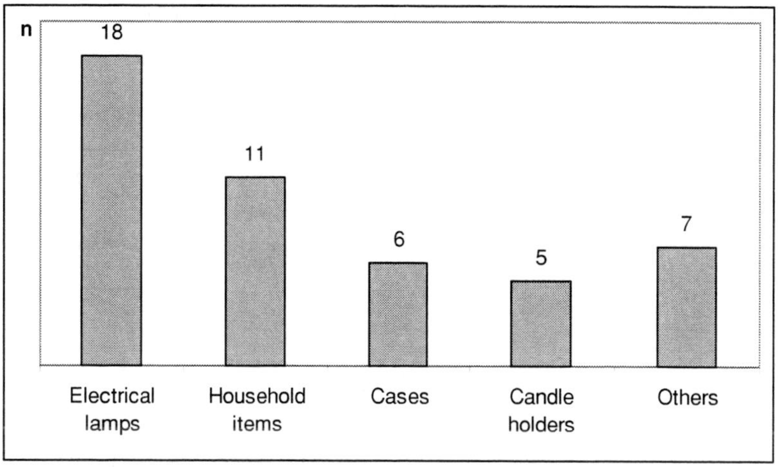

Graphic 12. Most important handicraft products in terms of share of the annual sales volume

Table 44 reveals that 15 out of the 18 lamp producers are not registered, the non-lamp producers account for a similar ratio of 27 out of 30 enterprises. There is no correlation between the annual sales volume of the handicraft producers and the product innovation. Of 48 handicraft producing enterprises 37 do not even know their sales volumes. However, it seems that the lamp producers are better organised: out of the 11 enterprises, who know their sales volumes, 7 belong to the group of the lamp producers. The lamp producers employ more personnel than the others, who are mainly self-employers. On average the lamp producers employ 2,9 workers per enterprise, whereas the non-lamp producing employers only employ 1,2 workers per enterprise. If the self-employers are excluded, the averages are 4,4 employees per lamp producing enterprise and 3,3 employees per non-lamp producing enterprise. There were no correlations between innovation and machinery or the enterprise's duration in the Guadua business. However, the firms younger than four years and the firms older than ten years are more likely to produce lamps.

Table 45. Correlations between innovation behaviour and micro-economic features of the handicraft producers in 2002

SME Group	Formality/ Innovation	Sales Volume/ Innovation	Number of Employees/ Innovation	Machinery Equipment/ Innovation	Duration in Business/ Innovation
Handicraft (N=48)	No correlation. Only 3 lamp producers are registered	No clear correlation. However, 2/3 of the lamp producers know their sales volume, whereas only 1/5 of the non-producers do	More than 50% of the employers, but only 25% of the self-employers (no employees) are lamp producers	No correlation or difference between lamp producers and non lamp producers	No clear correlation or difference between lamp producers and non lamp producers
Furniture (N=12)	All the technologically advanced furniture producers are registered. Of the artisan producers only 50% are registered	The technologically advanced enterprises are also the ones with the highest annual sales volumes	On average the technologically advanced furniture producers employ about 12 workers, the others only 2.	Clear correlation between technological progress and possession of machinery	The artisan enterprises exist longer than the industrial ones. There is no clear correlation though

8.2 State of SMEs in 2002 and 2004

Of the 60 enterprises, which were investigated in 2002, 42 could be visited again in 2004. Eighteen enterprises out of the original total could not be located in the study region, nor could they be found in other places (see *graphic 13*).

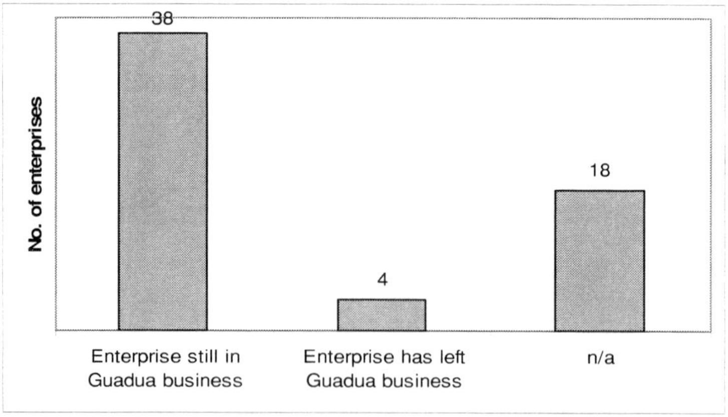

Graphic 13. Number of 2002-enterprises still working in the study region's Guadua sector

The four enterprises, which gave up the Guadua-business, did so because the sales and the business in general were unsatisfactory. These enterprises participated in the Guadua sector between two and nine years. Of the 18 enterprises, which could not be interviewed again, 10 had only participated in the sector for two years or less, three had spent between three and ten years in the sector and four had spent more than 10 years in the Guadua sector. In 14 of the cases investigations about the whereabouts of these enterprises revealed that the enterprises had moved to another region or the enterprise had disappeared without a trace[643].

Table 46. Exits of businesses, remaining enterprises and n/a-enterprises between 2002 and 2004

Duration in business in 2002	Number interviewed in 2002 (% of total)	Number still in Guadua- sector in 2004 (% of 2002)	Number of exits until 2004 (% of 2002)	Number of n/a[644] in 2004 (% of 2002)
< 2 years	25 (42%)	14 (56%)	1 (4%)	10 (40%)
3-10 years	21 (34%)	13 (70%)	3 (10%)	5 (20%)
11-20 years	9 (16%)	8 (89%)	-	1 (11%)
> 20 years	5 (8%)	3 (60%)	-	2 (40%)
Total	60 (100%)	38 (66%)	4 (5%)	18 (29%)

Analysing the age structure of the enterprises of the 2002 research campaign, it becomes obvious, that there is no interrelation between the age of the enterprise and its disappearance/exit. The assumption that especially the young firms (which comprise 50% of the interviewed firms) are more likely to leave the sector after a short time again, could not be verified. It seems that young firms probably change the location of their business more easily (40% mobility/disappearance). Nevertheless, the very old firms also account for a high percentage of mobility/disappearance (40% of the enterprises older than 20 years) (Both figures are listed in *table 45*; column Number n/a).

Summing up the number of firm exits and the number of n/a firms, the number amounts to 34% of the total examined in 2002, which means that 1/3 of the enterprises in the study region left the sector or at least the study region in 2003. Hence, 38 out of the originally 60 enterprises could be investigated in 2004 regarding changes in enterprise structure, products and production since 2002. The results are presented in the following graphics and tables.

[643] In only four cases the enterprise owner could be localised and briefly questioned about the present state of the enterprise. All the four entrepreneurs (duration in business between 1 and 15 years) said they had left the Guadua sector due to the bad sales and market situation.

[644] The enterprises, which could not be located anymore, belong to this group. However, it is likely that they continue to work in the Guadua sector but outside of the study region, as was indicated by neighbours or colleagues.

Graphic 14. Employment trends in the Guadua enterprises between 2002 and 2004 (n=38)

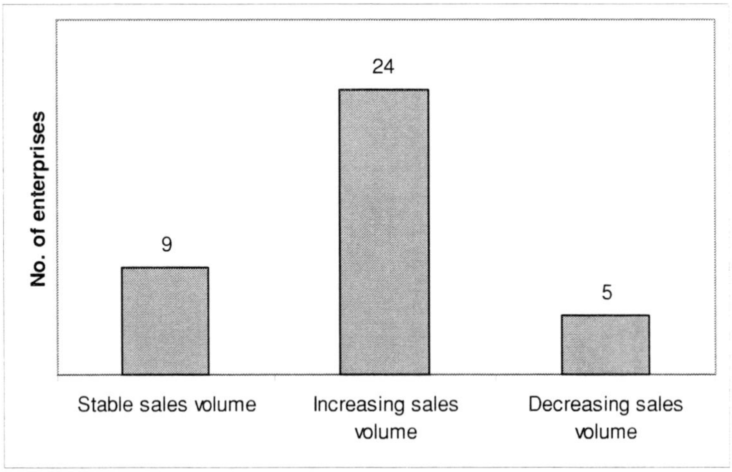

Graphic 15. Sales volume trends in Guadua enterprises between 2002 and 2004 (n=38)

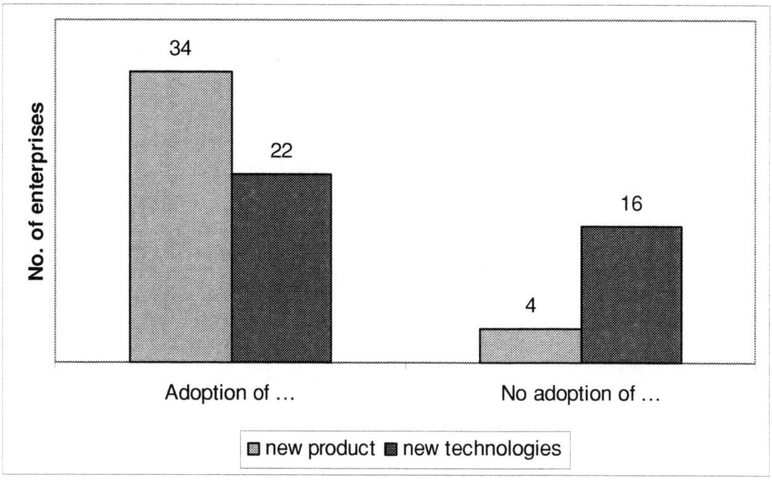

Graphic 16. Adoption of new products and new technologies in the enterprises between 2002 and 2004 (n=38)

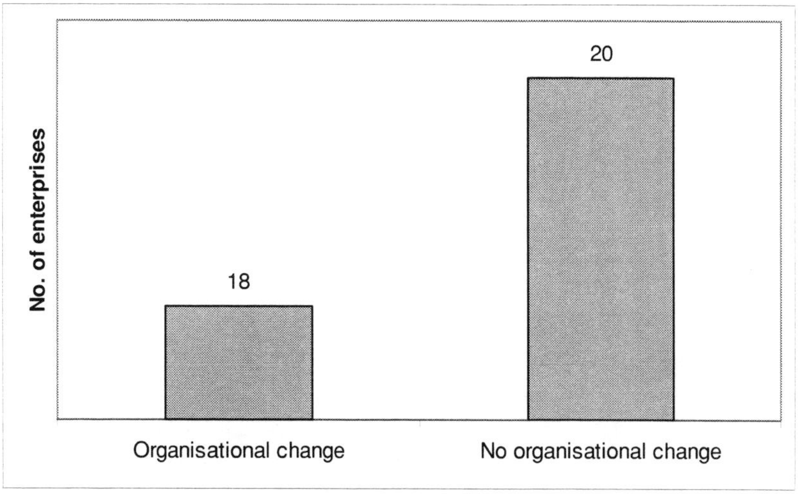

Graphic 17. Organisational changes in the enterprises between 2002 and 2004 (n=38)

Graphics 14 - 17 show, that the number of employees is stable or increasing and the annual sales volume has increased in the majority of the firms in between 2002 and 2004. Acquirement of new technologies (machines) and the adoption of new products by the enterprises are very common and reveal that the majority of the enterprises in the Guadua sector are motivated to invest and to adopt new technologies and products. Organisational changes were only implemented by a minority of the examined enterprises. In *table 46*, these results are broken down according to their relevance for the three types of SMEs: innovative handicraft producers, other handicraft producers and furniture producers.

Table 47. Comparison of innovators vs. others: Structural data from the Guadua sector

SME group investigated in 2002	Firm exits and n/a	No. Investi-gated in 2004	Increasing number of employees (% of No. investigated in 2004)	Increasing sales volume (% of No. investigated in 2004)	Adoption of new products (% of No. investigated in 2004)	Purchase of new machinery (% of No. investigated in 2004)	Organisa-tional changes (% of No. investigated in 2004)
Handicraft Innovators (n=18)	8 (44% of 18)	10	5 (50%)	6 (60%)	9 (90%)	5 (50%)	6 (60%)
Other Handicraft Producers (n=30)	10 (30% of 30)	20	3 (15%)	14 (70%)	19 (95%)	13 (65%)	8 (40%)
Artisan Furniture Producers (n=5)	1 (20% of 5)	4	1 (25%)	2 (50%)	4 (100%)	3 (75%)	2 (50%)
Industrial Furniture Producers (n=7)	3 (43% of 7)	4	2 (50%)	2 (50%)	2 (50%)	1 (25%)	2 (50%)
Total n=60	22 (37% of 60)	38	11 (29%)	24 (63%)	34 (90%)	22 (53%)	20 (53%)

Comparing the groups of enterprises, the comparatively high number of n/a belonging to the innovators` group reveals the weakness of the results in the *table 46*. However, basic trends in the enterprises could be detected: The innovators seem to increase the number of employees faster than the other entrepreneurs (the 2002 research already showed that innovators basically employ more people than the non-innovators). Second, the innovators are more likely to change the enterprise's organisational structures. On the other hand, the entrepreneurs from the group of the non-innovators seem to be more interested in investments in new machinery. *Table 47* highlights the differences in technological equipment of the three groups of enterprises (handicraft-innovators, non-innovative handicraft producers and furniture producers) and the major changes, which took place between 2002 and 2004.

Table 48.Technological progress of micro-level producers between 2002 and 2004

SME Group 2002	Changes 2004	Innovation status*
Product innovators (handicraft)	Basic wood processing equipment (saws, planes, ...) Industrial production process	improving lamps and shift to furniture sector (i and r)
Non-innovators (handicraft)	Manual equipment and wood processing equipment	Lamps as product innovation (r)
Furniture producers	Small complementary equipment	New furniture and new designs (i)

*i = incremental innovation; r = radical innovation

In *table 47* evidence of the validity of the theoretical approaches of 'innovation paths' and 'path dependency' is given. The production of lamps becomes more and more important to the group of handicraft producers, who have not produced them so far. Meanwhile the original lamp producers are moving on to other products with significantly more added value per production unit, namely furniture. For only five of the ten innovators in 2004, the lamps are still the most important product. However, for five of the ten new lamp producers lamps are the most important product now. In terms of process innovations the lamp producers seem to be more active investing in machinery for (pre)industrial production lines. In contrast the 'non-innovators' add special tools and machinery to their equipment (e.g. for the production of lamps).

Already established furniture producers did not invest in new large-scale machinery, since they invested in their production lines in the previous years. The innovation paths of the SMEs and the respective products are shown in *flow-diagram 1*. It can be seen that the fluctuation and the total number of enterprises in the early product stages is high. In these early stages of the enterprises and the product development the rate of informal business structures is high until they change from handicraft to furniture production or invest in large equipment.

The organisational changes, which accompany the structural and productive changes in the firms, are changes in the management structure (usually a new business partner) and changes in the marketing strategy (new outlets and new clients). The non-innovators do not differ significantly from the innovators when changing their organisational patterns. In all but one case, radical product innovations were accompanied by organisational changes (nine cases of new lamp producers and four cases of new furniture producers).

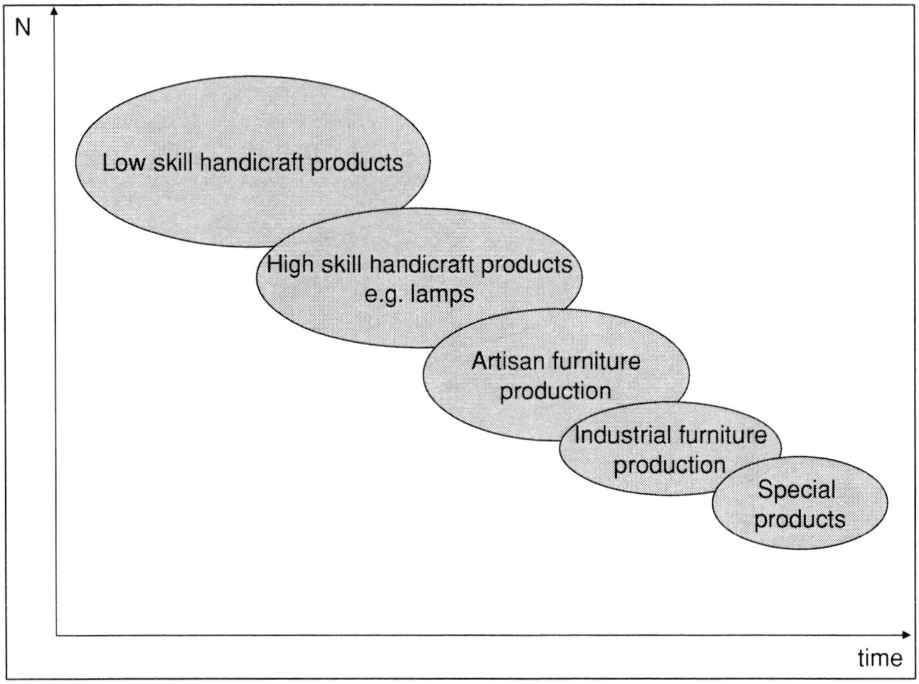

Flow-diagram 1. Innovation paths of handicraft and furniture SMEs in the Colombian Guadua sector

8.3 Interactions of SMEs

According to the theory, besides the active innovation oriented R&D activities, there are various ways of how knowledge is acquired within a group of enterprises. Generally, knowledge about products and production processes (as well as knowledge about markets and marketing) can be classified as individual, tacit knowledge of the entrepreneur and as the public, codified knowledge of the society/sector. This knowledge is spread within an economic sector according to three types of diffusion processes: 1) knowledge spill-over (externalities), 2) extension and education and 3) co-operation. Both the type of knowledge and the type of diffusion play an important role in the technological development of economic sectors, and the combination of both define the rationality of each entrepreneur, which in turn determines his/her entrepreneurial decision-making.

The research hypothesis that interactions, either in form of active inter-entrepreneurial interaction (co-operations), passive interaction between entrepreneurs (spill-overs) or interactions between the meso-level and entrepreneurs (through E&E and I&I) are the major driving force behind the innovation behaviour of micro-level entrepreneurs is analysed in the following sections of this chapter. A fourth type of interaction is the market and its clients, who may encourage innovations through their demands and preferences[645]. To test these assumptions, the innovation itself (for example the production of lamps or the introduction of new machinery) is understood as a success indicator for interactions with other actors at the micro-level and the meso-level. *Table 48* presents the answers of the 20 lamp producers about the origin of their knowledge on the production of Guadua lamps.

Table 49. Source of knowledge for lamp production

	Other handicraft producer (Co-operation)	Meso-level extension and education activities (E&E)	Autodidactic/ Copying and trial and error (Spill-overs)
Lamp producers 2002 (n=10)	2	2	6
New lamp producers 2004 (n=10)	1	3	6

When a new product is adopted or the enterprise itself is undertaking R&D activities to develop new products and processes, the enterprise has expenditures. When asked about adoption/R&D processes, 15 of the 30 handicraft producers interviewed in 2004 stated that they were engaged in such activities in the previous year. These enterprises perceive all the newly adopted or invented products as successful in the market and conclude they are worth producing. Thus, it seems that product failures are not common in the sector or are not perceived as such. The most important problems during the development process of the product were 1) related to the raw material (inadequately preserved or insufficiently dried Guadua) and 2) unavailability of adequate materials and finishing products, such as glues and varnishes. The entrepreneurs did not mention problems or obstacles because of legislation, product regulations, norms and other institutions. *Table 49* lists the most important invested resources during the trial and error process to adopt or develop a new product.

Table 50. Invested resources in the trial and error process (n=15)

Type of costs	No. of entrepreneurs
Entrepreneur's time	9
Experiments with product finishings	5
Labour costs	5
Adjusting and modifying machinery	3
Raw material treatment (Guadua)	3
Problems with providers of raw materials	1

645 See *chapter 3* on the theoretical approaches on innovations (technology-pull and demand-driven) and diffusion.

8.3.1 Interactions between Guadua based SMEs

In the Guadua sector the most common form of inter-entrepreneurial interaction as a source of knowledge is the spill-over (*Table 48*), i.e. no active distribution of knowledge but passive learning and copying processes resulting in entrepreneurial research based on 'trial and error'. The freely available knowledge (externalities of other entrepreneurs and their research) in the sector forms a sector pool of knowledge. As shown in *table 48*, copying other producers is the most important path of acquiring knowledge. Those who copy make use of the technical and creative knowledge externalities, which the original producers provide through the sale of their products (which was originally tacit, individual knowledge).

All the lamp producers (new and old) said, they changed and modified the lamp models they had originally seen or been taught to produce. They are adding tacit knowledge to the copied product. The list below shows the types of modifications, which the handicraft producers applied to their products. As *table 50* shows, technological modifications (e.g. electrical equipment) are not common amongst the group of interviewed handicraft producers (neither the old nor the new ones).

Table 51. Modification of lamps

Type of modification	No. of lamp producers (n=20[646])
Design and visual changes	19
Quality of the product (materials and preservation)	20
Technological modifications	1

The general degree of inter-entrepreneurial co-operations amongst SMEs in the sector is low. The sector is dominated by lose interactions between the enterprises. *Table 51* illustrates that business co-operations, such as joint purchase of raw material, joint marketing or production schemes are more frequent amongst enterprises, which are not producing lamps, than amongst the lamp and furniture producers. Lose business co-operations are common amongst all SMEs, i.e. the mutual exchange of services, such as transport, manual machinery or tools or the use of machinery in workshops.

[646] 10 lamp producers of the 18 visited in 2002 plus 10 producers of lamps, who started in 2003.

Table 52. Inter-entrepreneurial interactions

SME Group	Close Business Co-operations	Lose Interactions		
		Lending Materials	Exchange of Services	Exchange of Experiences
Lamp producers (n=18)	5	0	14	0
Other Handicraft Producers (n=30)	20	1	20	2
Artisan Furniture Producers (n=5)	1	0	3	1
Industrial Furniture Producers (n=7)	3	0	6	0
Total (n=60)	29	1	43	3

Table 51 underlines the results of *table 48*. The dominant reason for the adoption of new products is the spill over effects rather than the direct co-operation between the entrepreneurs. Only, the industrial furniture producers seem to look for inter-entrepreneurial co-operations more intensely than the other SME groups. A further assumption of innovation research is that knowledge is spread by the fluctuations in the labour market. This diffusion and development of knowledge through the labour market and fluctuation of employees could not be proven. However, since the innovative enterprises employ more staff than others (see *table 44*), it is more likely, that these employees use and modify the knowledge acquired during their employment. Only two lamp producers know of former employees producing and selling Guadua lamps in their own business.

8.3.2 Interactions between SMEs and the Meso-level

As *table 48* indicates, the general level of interactions between the micro- and the meso-level is not very high. Nevertheless, almost all the investigated entrepreneurs participated at least once in a educational workshop or seminar about Guadua offered by governmental organisations or NGOs. The interaction between the meso- and the micro-level actors are highlighted in *table 53*.

The active integration of micro-level actors into the meso-level's R&D and institutions-building processes is mainly reserved for large-scale enterprises of end products (*Chapters 8.4*). The focus of E&E activities and the promotional projects is on SMEs at the micro-level, as described in *chapter 7*. In *table 48* it is illustrated that only 5 of the 20 lamp producers had acquired their technical knowledge about lamp production through E&E activities. In *table 53*, the interactions between the meso- and the micro-level are listed. These could be identified during the interviews with the 60 handicraft and furniture producers in 2002.

None of the interviewed SMEs was integrated into meso-level R&D activities and none of them was involved in institutions-building. The major interactions of the interviewed SMEs with meso-level actors were based on the participation in E&E activities, which all of the investigated SMEs did at least once. About 25% of the 60 enterprises participated in SME promoting projects. The group of 'non-innovators' accounted for a larger proportion than the other two SME groups. There is virtually no access to credits for all types of SMEs. With respect to the legal institutional framework most of the questioned entrepreneurs were not aware of any influences, which laws and regulations may have on their enterprises (more than 50%). However, there are more SME entrepreneurs, who currently perceive the impact of the legislation as business promotion rather than as an obstacle. No direct correlations could be found between norms or resource management regulations and the innovative behaviour of the SMEs. A correlation between innovation and evaluation of the institutions could not be detected during the data analysis.

The institutional arrangements, which influence the Guadua sector's SMEs, are perceived positively across all SME groups (*Table 53*). This shows that the positive attitude towards the legal framework was predominantly expressed by unregistered handicraft SMEs. The registered handicraft SMEs were more reserved about the role of the formal institutions and had a rather neutral attitude towards them (no effects on their business). In comparison more entrepreneurs of the furniture producing SMEs perceive the institutional framework as positive. For the correlations between the registration status and legal regulations see *table 52*.

Table 53. Formality and perceptions of Guadua related institutions affecting SMEs

SME Group	*Institutions promotes my Guadua business*	*Institutionss are an obstacle to my Guadua business*	*Institutions do not effect my Guadua business*
Formal Handicraft SMEs (n=6)	2	1	3
Informal Handicraft SMEs (n=42)	9	4	27
Formal Furniture SMEs (n=5)	3	0	2
Informal Furniture SMEs (n=7)	4	2	1

Table 54. Innovation and interactions between micro- and meso-level actors in 2002

SME Group	No. of enterprises, which participated in E&E activities (% of n)	No. of enterprises, which participated in SME promotion projects (% of n)	No. of enterprises, which had access to credits (% of n)	No. of enterprises in regular contact with law enforcing entities[647] (% of n)	The effects of laws and regulations on the Guadua enterprise are perceived as...(% of n)		
					...hindering development of the enterprise	*...promoting the development of the enterprise*	*No such laws or regulations are known or they do not affect the enterprise*
Innovative Handicraft Producers (n=18)	18 (100%)	4 (22%)	0 (0%)	3 (17%)	2 (11%)	5 (28%)	11 (61%)
Other Handicraft Producers (n=30)	30 (100%)	9 (30%)	2 (7%)	8 (27%)	3 (10%)	6 (11%)	21 (70%)
Artisan Furniture Producers (n=5)	5 (100%)	1 (20%)	0 (0%)	1 (20%)	1 (20%)	3 (60)	1 (20%)
Industrial Furniture Producers (n=7)	7 (100%)	2 (29%)	0 (0%)	5 (71%)	1 (14%)	4 (57%)	2 (29%)
Total (n=60)	60	16	2	17	7	18	35

8.3.3 Intercations of SMEs with Clients and Suppliers

Innovation research typically emphasises the influence of the market demands on the entrepreneurs and their product decisions. Thus, the market related interactions of the Guadua SMEs were also investigated within the scope of this study. The SMEs were analysed with regard to the structural organisation of their buyers and their sales. In *table 54*, these structures are presented for three types of SMEs: innovative lamp producers, other handicraft producers and furniture producers.

The common (non-innovative) handicraft producers seem to rely more on regular buyers than the other SME groups. Entrepreneurs from all types of SMEs have occasional customers. For all SME entrepreneurs, sales to intermediaries, who sell the products on to other departments and the large cities such as Bogotá or Medellin, are of special importance. A third group of buyers are other handicraft and furniture producers. They are usually wholesalers of any type

[647] These law enforcing entities are the regional environmental authorities (CARs) and the DIAN (Dirección de Impuestos y Aduanas nacionales de Colombia), which is the national tax controlling entity, as well as the chambers of commerce and trade.

of these products and have larger shops or showrooms in the cities. To them the Guadua products are only a niche product. The handicraft and furniture producers frequently stated that the private costumers are more demanding in terms of design and creativity. The intermediaries and wholesalers put emphasis on the basic quality of the products and the reliability of the producer in producing the ordered quantities. However, the intermediaries and wholesalers also travel through the study region like hawkers and frequently buy their products coincidently.

The SMEs were asked about the regularity of important sales. To the enterprises important sales are determined by the sold quantity. The enterprise must operate at full capacity to produce the required quantity. The quantity itself depends on the size of the enterprise. Only few of the SMEs expressed that they had important sales, but if so, the orders were made by intermediaries. Across all SMEs, the larger enterprises were the ones, which had important sales regularly, which is based on mutual business interests between the intermediaries and the enterprises.

Table 55. Client and sales structure of Guadua SMEs in 2002

SME group	**Number of SMEs selling to ...**				**Number of SMEs having regular important sales ordered by...**		
	...permanent clients **(% of n)**	***...occasional clients*** **(% of n)**	***...other handicraft/ furniture producers*** **(% of n)**	***...intermediary dealers*** **(% of n)**	***...private clients*** **(% of n)**	***...intermediaries*** **(% of n)**	***...the tourist sector*** **(% of n)**
Innovative Handicraft Producers (n=18)	7 (39%)	10 (56%)	1 (6%)	14 (78%)	3 (17%)	7 (39%)	2 (11%)
Other Handicraft Producers (n=30)	15 (50%)	19 (63%)	3 (10%)	18 (60%)	3 (10%)	8 (27%)	3 (10%)
Artisan Furniture Producers (n=5)	0 (0%)	1 (20%)	5 (100%)	5 (100%)	0 (0%)	0 (0%)	1 (20%)
Industrial Furniture Producers (n=7)	4 (57%)	6 (86%)	7 (100%)	6 (86%)	0 (0%)	3 (43%)	1 (14%)
Total (n=60)	26 (43%)	36 (60%)	16 (27%)	43 (72%)	6 (10%)	18 (30%)	7 (12%)

Besides the impact the clients of the SMEs may have on the products and production processes, the interactions and co-operations with their suppliers of raw materials and semi-fabricated products are seen as potential sources of innovations. *Table 55* presents the co-operations of the Guadua SMEs with their suppliers.

Table 56. Co-operations with providers of raw material according to SME group

	Innovative Handicraft producers (n=18)	Other Handicraft Producers (n=30)	Artisan Furniture Producers (n=5)	Industrial Furniture Producers (n=7)
Co-operations with Guadueros	0	2	1	0
Co-operations with depots	0	0	1	0
Co-operations with other furniture and handicraft producers	3	9	0	0
Co-operations with farm owners	4	3	2	4
Total and % of n	7 (39%)	14 (47%)	4 (80%)	4 (57%)

The number of co-operations between the raw material suppliers and the SMEs is not high. The common handicraft producers are more engaged in such relations than the other SME groups. Frequently, these SMEs have relations to farm owners, owners of depots and other larger SMEs, where they can obtain smaller quantities of such raw material for free or at low prices. In many cases the raw material, which the small SMEs use, was disposed of by larger SMEs, or due to certain shapes of the Guadua canes the material is not suitable for construction purposes. The larger, more industrial SMEs in this study either have their own Guadua stands or they maintain close relationships to Guadua farmers. They may also maintain contacts with Guadueros, who know the special needs of the particular enterprise and provide the required raw material in larger quantities.

8.4 Special Providers of Guadua Products

The traditional commercialisation chain of Guadua in the study region is the chain from private small and medium size farmers via the independent working *Guaduero* via the retailing depots[648] to the final private consumer in the construction sector (see *chapter 4*). HELD AND MANZANO (2004) estimated an average annual sales volume of approx. US$ 1.3 Mio. in this chain annually. Apart from this traditional supply chain of the construction sector, large scale producers of Guadua raw material have increasingly entered the market in the last five years. They have different types of clients, usually architects and civil engineers.

The large scale producers do not only provide basic Guadua products, but also meet the specific demands of architects and engineers. In any way they offer close co-operation to their clients, including an intensive exchange of knowledge and experiences between the constructors and the raw material providers[649]. Generally, these large-scale producers have access to large amounts of the Guadua resources or Guadua raw materials and conduct a registered agricultural/agro-industrial business. In *table 56* the most important features and the market segments of these enterprises are described.

Table 57. Market orientation of special Guadua providers

Market / Product type	Construction (large scale, housing)	Construction (large scale, artistic)	Handicraft and Furniture	Export Markets
Standard material, assorted and treated (Borax)	No. 1; No. 6	No. 1	No. 1	No. 1
Standard material, assorted and treated (Fumigation)	No. 5, No. 6	No. 1, No. 5	No. 5	No. 5
Standard material, assorted and treated (Diesel)	No. 3	No. 3		No. 3
Standard material, assorted, treated and considering phases of the moon	No. 1	No. 1	No. 1	
Standard material, assorted and untreated	No. 2, No. 6	No. 2		No. 2
Special assortments, treated (Injection)			No. 4	
Pre-processed products (laths and laminated bamboo)		No. 4, No. 6	No. 4	

(Each number represents one interviewed producers)

648 This step may be left out in case the Guaduero is delivering directly to the construction sites.

649 Here is the point, where the assumption that close producer-client co-operations promote innovations is probably the most verified in the present study.

During the fieldwork in 2002 the strategic production schemes of six special providers were developed (see *table 56*). The strategies of these actors aim at avoiding competition by occupying niches of the market. In 2003/04 the products and the business organisation of most of the investigated enterprises changed as shown in *table 57*.

Table 58. Changes in business orientation of special providers in 2004

	Market	Production capacity	Production line	Business organisation
No. 1	Increasing national and international demand	Amplification	Preservation capacity extended; lath production possible	Participation in business cartel
No. 2	Increasing focus on international markets; increasing prices in national market	Unchanged	Preservation (Borax) line planned; lath and parquet production planned	Cooperation with national and international entrepreneurs
No. 3	Unchanged	Amplification	Borax preservation offered	Unchanged; better contacts to pre-providers
No. 4	Reorientation towards new products and consumers; Artisan production eliminated	Amplification of lath production	Strip and laminated bamboo production line established; Borax soaking and injection preservation offered; New drying facilities established	New business partner
No. 5	Unchanged	Unchanged	Unchanged	Business guided by manager
No. 6	Out of business	Out of business	Out of business	Out of business

(Each number represents one of the interviewed producers)

In 2002 the market leader (enterprise No.1) occupied a special segment of the market. Although the enterprise, which was established in 2001/02, had problems in that year to maintain the liquidity of the firm, it could establish business contacts and more importantly, a high quality reputation of its products. Losses and capital investments of the first year were balanced by other divisions of this agricultural enterprise. The success of this firm in the year 2003 and the fact, that the firm could establish the market prices, motivated other providers to enter this market segment. This is accompanied by the notion, that enterprise No.1 is temporarily not able to satisfy the current demand. Other important factors are that the No. 1 firm successfully established a new production line and its products have been accepted by the market. For the other entrepreneurs this enterprise is holding a role model position. For them, the risks of taking up this type of production line and marketing strategy on large scale are reduced. The No.1 provider is freely providing information about production technologies, failures, as well as research and development processes. Market information is dealt with more carefully.

The other firms in the sector are pursuing a strategy of adoption and amplification. They seek to achieve at least a minimal level of equivalent quality standards. None of the firms has yet aimed at producing further developed end-products on their own. They see their strength in mass production of raw materials of various qualities and the production of semi-manufactured products for further use in the wood processing industries.

According to the special providers, new technologies and products are considered to be the key to enter international markets. National markets are considered to be markets of low-value products without significant profits. Since new technologies usually exceeded the capabilities of the original firm owners (usually coffee growers and cattle breeders), new business partners were chosen according to their background knowledge on production processes (e.g. from the wood processing industries) and experiences in markets. The contacts to those persons were established in different ways. No typical strategy of choice of partners could be identified. The entrepreneurs articulated that they would prefer to work with people they already know in case they provide the required expertise.

Under the special considerations of investment management it has become obvious, that investments are undertaken based on private sources of capital. None of the examined enterprises acquired foreign capital from financing organizations or schemes. In case of investments, the capital used, originated from own sources (private capital) and/or the number of business partner was increased. Except for one, the entrepreneurs generate income from various sources (mainly from agricultural production). Initial losses of the Guadua related enterprises can be balanced through positive business results of the other branches of the whole enterprise. The usual legal form of the enterprises is the S.A./S.L. (*Sociedad anonima/Sociedad limitada*), which offers tax advantages and the liability is restricted to the firm capital[650].

During the phase of trial and error in the product development cycle, sunk costs arise. These costs are usually not considered in the overall calculation of the business, although the entrepreneurs try to minimise them. This is attempted by using farm employees and infrastructure and machinery already utilised by the main business. Time and transaction costs of the entrepreneur are considered to be of much higher importance, such as negotiations with new partners, new clients, marketing and bureaucracy.

The selected providers are better integrated into the processes of institutions-building and communicate more with the meso-level actors than the SMEs. Due to their status as large-scale agricultural producers of various products, they are registered enterprises. They are owners of large stands of natural Guadua resources and some plantations. Hence, they are constantly in contact with the regional environmental authorities, the secretary of agriculture and the DIAN. Nearly all of the interviewed providers complained about the regulations, which apply to the utilisation of the resources. According to them, the regulations are too restrictive and not adapted to the potentials the Guadua resources are offering. Moreover, they criticised the lacking institutional support with respect to international trade and export. In their opinion there is too much bureaucracy and the transaction costs hinder national and international trade with Guadua raw materials. Due to the perceived deficits in the existing

[650] CAMARA DE COMERCIO DE BOGOTA (2002)

institutions and because of the registration status and professionalism of their Guadua enterprises, they actively seek to influence decision-making and the institutions-building processes.

The special providers interact with the meso-level actors during joint R&D activities, mainly silvicultural on-farm research. There is also an exchange of experiences with Guadua processing machinery and production lines. The providers are also demanded instructors for E&E-activities conducted by the meso-level actors, and they are frequently invited as speakers at seminars, workshops and conferences. In *table 58* their integration into the processes of institutions-building and other co-operations with meso-level actors are presented.

Table 59. Interactions of special providers with the Guadua sector's meso-level actors

Special Provider	Involvement in institutions building [651]			Co-operation with meso-level actors in R&D	Instructor or speaker at meso-level E&E activities	Leading member in a meso-level organisation
	MSBC	*ICONTEC*	*Guadua Council*			
No. 1			X	X	X	
No. 2		X		X	X	X
No. 3			X		X	X
No. 4		X	X		X	
No. 5				X		
No. 6[652]	-	-	-	-	-	-

In the study region three new special providers of Guadua products appeared in 2003. At the moment their respective markets are very fragile and they strongly depend on only one or a few clients. The newcomers make use of the latest technologies (lath saws, kiln drying, bouchery preservation). Furthermore, in all the three cases knowledge about technology and processing is provided by outsiders and/or enterprise partners from the wood-processing sector. Characteristics of the newcomers are listed in *table 59*.

The two private enterprises (A and C) provided the starting capital through private capital contributions. The third, project based enterprise (B), has access to a low rate credit scheme. In two cases the emergence of new providers in the study region is closely related to meso-level activities in the field of Guadua sector promotion. They are 1) the spin-off of a research project and 2) project funding. The third newcomer did not benefit from any meso-level activities. Since all the three enterprises are still in its initial phase, indications on success or failure of entrepreneurial and meso-level decisions would be premature.

[651] See *chapter 7* for a detailed description of the processes.

[652] The special provider gave up Guadua related activities in 2003.

Table 60. Characteristics of new special providers

Characteristics	Enterprise A	Enterprise B	Enterprise C
Origin	Private-Research Partnership	Technical Co-operation Project	Private Initiative
Capital contributions	Private	Low rate credit and municipal support	Private
Products	Treated Guadua; Kiln-dried strips	Treated Guadua; Laths	Specially treated Guadua
Market access	Indirect; national and international	Indirect; national and international	Direct; international
Client structure	Wide; insecure demand	Wide; insecure demand	Narrow; secure demand
Know-How Input	Business partners, partially from the wood-processing sector	Outsiders, wood- and Guadua-processing sectors	Business partners from wood processing sector
Projection	Acceptance in FOMIPYME-programme	Positioning in local markets	Securing export market

9. Discussion

Starting from the hypotheses regarding the importance of innovation for the competitiveness of economic sectors (clusters), this study sought to answer a set of open questions, which have not yet been answered by theory and empirical research. The focus of the research was on forest-based sectors in developing countries, on the micro and the small and medium sized enterprises, and the actors at the meso-level.

The following discussion is divided into six sub-chapters: *Chapter 9.1* summarises the research results from *chapters 6 – 8*. *Chapter 9.2* compares the results of this reseacrn with studies from bamboo sectors in other countries. The following chapter (*9.3*) discusses these results in the context of the national Colombian econmy and its industrial sectors. The *chapters 9.4 and 9.5* addresses the theoretical assumptions outlined in *chapters 2 and 3* of this thesis. The last sub-chapter (*9.6*) contains conclusions regarding the implications of this study for theory and practice; it also suggests areas of further cluster and innovation research.

9.1 Summary of the Research Results

The research questions aimed at testing the economic growth theories with respect to the origin and function of innovations. The analytical framework system chosen was adopted to provide a better understanding of competitiveness in the context of economic sectors in developing countries, with an emphasis on economic and social aspects. This analytical system, 'systemic competitiveness', is based on the theoretical approach of the regional agglomeration of firms and their interactions. The case study presents the Colombian bamboo (Guadua) sector, an economic sector based upon the use of a forest resource in the rural areas of a developing country.

9.1.1 The Meso-level Analysis

The basic research questions underlying the analysis of the meso-level were:

- What are the meso-level actors' views about the present situation and the future development of the Guadua sector?
- How and why do they promote the Guadua sector? What are their strategies?
- How do the meso-level actors interact (co-operate/co-ordinate/compete) and what are the reasons for these interactions?
- Do the interactions between the meso-level actors determine their promotion activities?

The results presented were based upon personal interviews, consisting of a guideline and a structured interview part. Forty three interviews were conducted with meso-level actors from the Colombian Guadua sector in the year 2002; 14 complementary guideline interviews were conducted with further meso-level actors in 2004. The meso-level actors interviewed belonged to various institutional backgrounds, such as governmental organisations, NGOs, research entities, etc., and were chosen according to their activities in the Guadua sector or the impact they have on it.

The interest of meso-level actors in the Guadua sector has increased over the last five years. More than 50% of the meso-level actors interviewed began their Guadua sector related activities in 1998 or later (see *graphic 4*).

Two important impulses for this increasing interest have been: 1) the continuing crisis in the coffee sector, the major source of income in the study region; and 2) an earthquake, which hit the region in 1999 and revealed the competitive advantages (earthquake resistant traditional buildings) of Guadua-based construction and which demonstrated the application of Guadua constructions in social housing projects for the victims of the earthquake. The meso-level actors assume that the promotion of industrialisation processes in the Guadua sector will contribute to the socio-economic development of the study region, i.e. to counteract the negative effects of the coffee crisis and to offer employment, income and housing to fringe groups in the study region.

The predominant perceptions of the meso-level actors in relation to the Guadua sector are that there is a lack of cultural and economic validation of the resource base and its products, and that there are deficits in the institutional framework to promote sustainable resource use and national and international trade in Guadua products. These actors see problems in inter-organisational co-operation, communication and co-ordination (see *table 29*). Therefore, the actors have developed strategies to overcome the perceived problems in the sector, to pursue the aims they have fixed for the sector complying with their duties and self-understanding (*Table 31*).

To achieve their goals, the meso-level actors use a mix of activities (*Table 32*). These activity mixes can be described as an activity profile of each meso-level actor and structured into research and development (R&D) activities, education and extension (E&E) activities and institution building and interaction promoting (I&I) activities. While certain actor groups, such as the governmental actors, mainly focus in I&I processes and activities, others such as public-private actors and the research entities mainly conduct R&D activities. The individual activity mix profile of all actors changed between the period before 1998 and in the years thereafter. (See *net diagrams 1 to 6*)

According to the analysis of the meso-level actor interviews, participation in the institution building processes constitutes their main Guadua sector related activity. In these processes, the governmental organisations hold a leading position. These institution building processes are: 1) the formulation of sustainable use and management regulations by the Ministry for the

Environment and the regional environmental authorities under the framework of the MSBC[653]-program; 2) establishing product and production standards and codes guided by the Colombian standardisation and normalisation institute ICONTEC; and 3) the formation of a Guadua production chain within the framework of Colombian agricultural legislation. The MSBC process has been conducted mainly without the integration of micro-level actors and has focused on the harmonisation of the various use and management regulations for bamboo resources in the study region. This resulted in the elaboration of common use regulations for the departments of the study region in the years 2001 and 2002. The micro-level actors perceive these use regulations as too restrictive and protectionist. The MSBC is currently working on amendments to their regulations to adapt them more to the needs of the micro-level.

In the ICONTEC process, the actors are working on the elaboration of product and production standards based on international ISO standards and certification. This process comprises entrepreneurs, researchers, governmental actors and NGOs. Some meso-level actors and entrepreneurs have stated that this process is starting too early, and will probably result in either 'weak standards', or alternatively the determination of standards common SMEs cannot comply with. The aim of the process is to elaborate proposals to be submitted to the national standardisation organisation (ICONTEC). Two proposals have been completed to date, both on the harvesting and treatment of raw materials.

The third institution building process is guided by the national Guadua council, which is thought to be the constituting entity of a Guadua productive chain under Colombian agricultural law. This process has just started and an evaluation is not yet possible. However, all relevant actors have been considered so far, but the individual roles have not yet been clarified.

Of the three institution building processes, the MSBC-process is the oldest and the institutions created have already been enforced. Its major success has been the harmonisation of various utilisation regulations in the four study departments. Although initially it was a top-down governmental process, the experiences of the first years of the utilisation and management regulations paved the way for closer co-operation between the meso and the micro-level actors. The ICONTEC process is integrating more micro-level actors and is an important cornerstone in relation to product quality standards and production processes, which in turn will contribute to the improved national and international competitiveness of the sector. However, R&D did not accompany the development of the quality standards. The standards fixed in the guidelines result from individual experiences and the results of only a small amount of systematic research. They seek to comply with international norms, which do not reflect the real production capacities of the majority of the micro-level entrepreneurs. The third process, the formation of a Guadua production chain, is important in terms of the organisation of the sector's actors and increasing recognition from the agricultural sector, as well as facilitating access to state support and funding. For the future of the sector this

[653] Spanish: *Manejo Sostenible de Bosques de Colombia*

organisation form is desirable. However, the Guadua sector is not a typical agricultural commodity-producing sector, but rather a mixture of agricultural producers, artisans and industrial manufacturers with consumers positioned in end use markets ranging from the construction sector to handicrafts.

An important theoretical assumption of this research is that interactions are an important driving force behind successful sector promotion strategies. This was tested through the analysis of the intra-actor interactions. As the dominant actor group, the governmental actors currently guide the activities in the sector. The highest degree of co-operation between the meso-level actors was found in R&D, with the public-private actors the outstanding actor group in terms of interaction ratio. The degree of co-operation between the meso-level actors in E&E activities is comparatively low. This can be explained by the fact that for several meso-level actors E&E activities are income-generating activities and, therefore, conducted individually. (See *graphics 5 and 6*)

There is a general notion that the interaction patterns between the meso-level actors are complex and uncoordinated, whereas their links to actors in other economic sectors or industries are uncommon. For example, no official co-operation between the forestry and wood processing sector and Guadua related meso-level actors could be identified (see *graphic 7 and venn-diagram 1)*. The actors interviewed represent all meso-level actors engaged in Guadua related activities, plus a limited number of meso-level actors, who keep a lower profile in Guadua related affairs, but also affect the sector indirectly through their activities. These actors are SME promoting NGOs and credit institutes. These entities perceive the Guadua sector as just one of many income-generating possibilities. In their opinion, the SMEs from the Guadua sector are not as competitive and innovative as other agro-industrial and industrial enterprises, and basically lack economic feasibility and perspectives (see *chapter 6.4.2*).

The meso-level actors themselves receive technical and infrastructural support, as well as public funds from the national research funding organisation COLCIENCIAS, which is the main organisation of the national Colombian innovation system. This support is geared towards the promotion of R&D and E&E activities. COLCIENCIAS has only funded ten Guadua related R&D and E&E projects in the last ten years (see *graphic 8*). According to the organisation this is due to the low number of research proposals submitted. The tendency is rising. COLCIENCIAS' major criticism regarding Guadua-related research proposals is the lack of economically feasible application possibilities for the expected research results[654]. COLCIENCIAS is increasingly funding SME- and innovation-oriented research projects, whereas Guadua related proposals are mainly for basic research projects.

[654] Compare COLOCIENCIAS`guide for the submission of reseacrh proposals. COLCIENCIAS (2002)

Conclusions

The meso-level actor perception of the Guadua sector's present situation and future perspectives was identified. The actors want to promote industrial development of the sector in order to contribute to regional development. Their institutional background and self-understanding, and the perception of the sector's status determine the manner in which they engage in promoting the Guadua sector.

All actors adopt various R&D, E&E and I&I activities, which reflect their access to technical, personal, financial and institutional resources. In general, governmental and public-private actors, involved in R&D and I&I, dominate these activities. They seek regulation of resource utilisation and the promotion of the industrialisation processes. Interactions between the meso-level actors are common, although the actors criticise deficits in inter-organisational co-operation, coordination and communication.

The manner in which the meso-level actors interact reflects the state of the Guadua sector. There is a lot of negotiation and interaction, but the outputs of these interactions are mainly seen as moderately, or in some cases even unsatisfactory. The meso-level actors view E&E activities, which communicate R&D and I&I to the micro-level actors, as sources of income. The R&D and I&I communicated through E&E usually represent the individual actor's experiences and perceptions. The Guadua related meso-level actors hardly maintain relations with other industries and economic sectors. Hence, an enrichment of knowledge from outside is not taking place. This in turn characterises the external image of the Guadua sector amongst research institutes, and SMEs promoting and credit lending entities, which perceive the Guadua sector as non-competitive and non-innovative.

9.1.2 Interactions between the Meso and the Micro-level

Answers were sought to the following research questions:

- How do meso-level actors integrate micro-level actors into their activities?
- How do the meso-level actors try to enhance the innovative capacity of the micro-level actors?
- Is it possible to draw conclusions from the interaction behaviour among the meso-level actors for the way they interact with the micro-level actors?

The *chapter 7* draws upon the analysis of the 57 meso-level interviews. Furthermore, it reflects the observations of various field trips to production sites of Guadua processing SMEs and projects. The institutional analysis was based upon secondary data research focusing on the legal framework for Guadua utilisation and trade in the study region, and in Colombia as a whole.

The analysis of the meso-level actor interaction strategies to influence the Guadua sector's micro-level revealed six interaction paths (*Table 35*): 1) the direct and 2) the indirect promotion of R&D in micro-level SMEs, 3) free E&E offers and 4) charged E&E activities, 5) I&I activities with the participation of micro-level actors, and finally 6) I&I activities, which are communicated to the micro-level, but do not involve its participation.

These meso-level actor interventions and interactions are based on their perceptions of the current state of the micro-level enterprises (*graphic 10*). According to the meso-level actors, the SMEs in the Guadua sector lack the technical knowledge to process the Guadua raw material properly, and furthermore, the SMEs do not generate sufficient added value. The meso-level actors perceive a lack of quality management in the sector, which results in weak national and international competitiveness. The major potential for innovation in the sector is expected to lie in the production of flooring for international markets, the use of Guadua as construction material for social housing programs and the services of Guadua for the environment (carbon sequestration and environmental services) (*graphic 11*).

Through its activities, the meso-level tries to assist the SMEs in overcoming deficits and to achieve its potential. The promotion of entrepreneurial R&D is based upon two strands: a) the provision of an infrastructure and consultation framework to support SMEs in upgrading and the industrialisation process, and b) the provision of access to micro and low rate credit schemes. The first is usually provided by a combination of governmental actors, public-private actors and NGOs. The credits are provided by NGOs and public and private banks. Both the NGOs and the banks select the SMEs they support according to certain criteria, of which economic feasibility is the most important. According to them, Guadua related enterprises rarely apply for credit, which they explain by the informal status of the majority of them, which limits their security and credit rating. The interviewed representatives (two NGOs and three banks) stated that they do not know of a single Guadua related enterprise currently receiving credit from their respective organisations. The criteria of the infrastructure and support providing entities comprise the innovativeness of a business concept and its potential in the market, as well as the long-term economic feasibility. According to these interviewees, no Guadua related project proposal has met the selection criteria so far.

Direct integration of micro-level entrepreneurs into meso-level R&D only occurred with the large-scale providers of Guadua products. These actors maintain close co-operation in form of on-farm and in-enterprise[655] R&D. Furthermore, these entrepreneurs are invited to speak at seminars and conferences, and to act as instructors during workshops organised by the meso-level actors. The SMEs in the micro-level do not receive this attention. They are perceived as clients, who participate in E&E activities. The recognition of the large scale Guadua providers may be explained by their valuable practical knowledge of the processing of raw Guadua, which is an important source of knowledge for the meso-level actors' own R&D-activities.

[655] The term in-enterprise R&D summarises the R&D meso-level actors conduct in co-operation with the entrepreneurs in their production facilities, such as workshops and production sites.

The E&E activities, which are thought to disseminate technical and entrepreneurial knowledge amongst the micro-level actors, mainly focus on handicrafts and construction using Guadua. These activities account for an increasing number of participants. There are two types of E&E activity; those free of charge (community workshops, extension and dissemination activities by NGOs and governmental actors) and those, which must be paid for (e.g. university seminars and workshops). Hence, the participants differ according to the type of activity offered: those participants willing and/or able to pay and those unwilling and/or incapable of paying for the E&E events. For a number of meso-level actors the income generating character of these activities hinders closer co-operation between them, but also ensures quality since the competition between the actors leads to the selection of experienced instructors and speakers. On the other hand, some meso-level actors engage in such activities without the necessary expertise and qualifications in Guadua related topics.

The third way in which the meso-level intervenes in the Guadua sector micro-level is through the I&I-activities, which comprise the institutions and the initiation of projects. Currently, there are three important institution-building processes in place. All three include some degree of participation by micro-level actors. They are less involved in the development of resource utilisation and management regulations (MSBC-process), whereas the development of product and production standards requires their participation (ICONTEC process). The third process is guided by the national Guadua council, which brings together actors from all Guadua backgrounds interested in the creation of a Guadua productive chain; a process resulting in high transaction costs for all actors (*Chapter 9.1.1*).

The promotion and creation of micro-projects and model enterprises constitute the sixth meso-micro interaction path. Four such Guadua related micro-projects were visited during the research campaign. The four projects were established within the framework of two promotion programs. One of them was especially created for the promotion of the utilisation of the Guadua resources, initiated by the MSBC program and the municipalities. The other program is not specifically directed towards the promotion of Guadua SMEs. However, two Guadua related projects have been approved for inclusion in the promotion scheme by the program. The first promotion program focuses in the promotion of fringe groups in communities of the study region and mainly promotes the production of Guadua raw material and manufactured products. Its aim is to establish community-based enterprises with a collective socio-economic impact. The second promotion program emphasises the promotion of individual SMEs within the structuring framework of supporting entities. A comparison of the major features of the four projects is presented in *tables 39a and 39b.*

Conclusions

The meso-level actors intervene in the micro-level through six different interaction paths of R&D, E&E and I&I activities, which seek to solve the perceived SME deficits and activate the sector's potential. The meso-level is integrating the micro-level actors (especially the special providers) into the institution building processes; to a greater extent in technical aspects, less so in the environmental ones. In terms of E&E activities, the micro-level entrepreneurs and farmers are seen as receivers or clients rather than as partners. Only selected micro-level actors are integrated in the meso-level R&D activities. The majority is only presented with the results of such activities later on.

The meso-level actors pursue the promotion of innovative behaviour on the part of entrepreneurs through the support of entrepreneurial R&D activities (credits and infrastructure support), through the communication of technical and entrepreneurial knowledge (E&E activities), through the establishment of utilisation and production standards and the creation of organisations to bring the actors together (I&I activities). The interaction behaviour, as analysed in *chapter 6*, influences the interaction paths of the meso-level with the micro-level actors. The lower the degree of meso-level actor interaction in a certain field, the lower is the integration of micro-level actors into these activities. Further, the lower the expertise of the meso-level actors, and the opportunity to acquire expertise from other meso-level actors, the higher is the integration of micro-level actors into meso-level activities. However, the micro-level SMEs and actors are perceived by the meso-level as technically underdeveloped and not economically feasible. To them, the micro-level consists of clients and receivers of knowledge and progress generated by the meso-level.

9.1.3 Innovations at the Micro-level

The *chapter 8* sought to clarify the following questions:

- How do SMEs in the Colombia Guadua sector innovate?
- Which influences have the interactions between SMEs on the innovative behaviour of the entrepreneurs?
- Which influences have the meso-level's promotion activities on the micro-level's innovations?

Sixty SMEs (handicrafts and furniture) from the Guadua sector micro-level were investigated in 2002. The same enterprises were sought out again in 2004, but not all of them could be located. Of the sixty investigated originally, 42 could be visited a second time. The focus of the investigation was on micro-economic features and their characteristic patterns of innovation and knowledge acquisition. Special attention was paid to the SMEs` interactions with the meso-level actors and their to the meso-level promotion strategies and interventions. In addition to the SMEs, six large-scale providers of Guadua raw material and manufactured products were visited and interviewed in 2002 and 2004. They were investigated with respect to their innovation and interaction management in comparison to the SMEs in the Guadua sector.

The research revealed that the majority of the investigated Guadua SMEs were not registered, or informal. No clear correlation between the formal/informal status of the handicraft SMEs and their micro-economic features could be identified. In the furniture sector, however, the size of the enterprise, its machinery and its sales volume correlated positively with formal status (*Table 43*). One consequence of the SMEs' informal status is that they cannot access credit or promotion schemes offered by the state.

The innovativeness of the SMEs was classified according to the OECD's definition of innovations in enterprises. For the handicraft producers the introduction to the market of a successful new product was determined an indicator of the innovativeness of the respective SME. For the furniture producers, the application of an 'advanced' production line was defined as an indicator of innovation. This resulted in a ratio of 18 innovative handicraft producers of 48 investigated, and 7 of 12 furniture producers investigated.

The informal status, a typical economic feature of the SMEs in the Guadua sector, correlated with other features, which were highlighted in *table 43*. The most important was the positive correlation between innovative furniture producers and formal status. The most important correlations between the innovativeness of the SMEs and other micro-economic features were the larger number of employees of innovative firms (handicraft and furniture), and the higher level of entrepreneurial skills of innovative entrepreneurs. Furthermore, the innovative furniture producing SMEs account for higher annual sales volumes (*Table 44*).

In the 2002 research campaign, it was established that more than 40% of the investigated enterprises had started only two years ago, or even more recently. In the 2004 follow-up the greatest proportion of inaccessible firms belonged to this latter group. The exit of these firms from the Guadua sector could not be confirmed definitively, but they had disappeared from their previous locations (*Table 45*). No correlations could be found between the age of an SME and its innovativeness. In the furniture sector, it was the artisans rather than the industrial SMEs who remained in the Guadua business longer, probably because the industrial ones change over to other types of raw material.

The major changes in the SMEs between 2002 and 2004 took place in terms of the adoption of new products and technologies, increasing sales volumes and a slightly increasing number of employees (*Graphics 14 - 17*). A tendency towards an increase in the number of employees was found especially amongst the innovative firms. The changes identified in the SMEs investigated underlined the theoretical assumption of the innovation paths, where through mainly incremental innovations, a sector's enterprises develop from artisan SMEs to industrial enterprises (*Flow-diagram 1*). In the case of the investigated SMEs, this innovation path followed the development line from artisan handicraft producers to industrial handicraft producers, who then started to produce furniture.

In terms of the source of knowledge for the adoption of new products and technologies, the investigated SMEs stated that the main source were the passive spill-over effects within the Guadua sector. Inter-entrepreneurial co-operation and the E&E offers of the meso-level were not of great importance to the entrepreneurs (*Table 48*). Very few of the investigated SMEs received a business related credit, and only few had participated in micro-project activities of the meso-level. However, all of them participated at least once in an extension and education event. The institutions governing the Guadua sector were perceived by the majority of the

SMEs as neither disturbing, nor promoting. However, as the majority of SMEs were not registered, they were not in a position to assess the full effects of the institutions. Nevertheless, the majority of the formal enterprises felt neither hindered nor promoted by the existing institutions. No clear correlations could be identified with respect to the influence of the enterprises' clients on the innovative behaviour of the Guadua SMEs (*Table 54*).

The large-scale providers of Guadua raw material and semi-fabricated products revealed different patterns of innovation and interaction behaviour compared to SMEs. The large-scale providers occupied niches in the Guadua markets and supplied materials treated and processed for use in manufacturing industries, as well as standard materials for the construction sector (*Table 56*). They were mostly large-scale farmers or had direct access to large areas of natural or planted Guadua resources. For the majority, the Guadua business was just one of several activities undertaken in the agricultural sector.

Between 2002 and 2004, one of these special providers gave up the Guadua business, whereas the others have extended their businesses and improved the manufacturing processes, as well as their market access (*Table 57*). Due to their ability to balance losses and initial investments in the Guadua business through other strands of their enterprises they have room to manoeuvre, as well as undertake experiments in their Guadua business. Through their R&D activities they provide important knowledge for the rest of the Guadua sector. They co-operate closely with their most important clients, i.e. skilled Guadua engineers and architects. Through this mutual interaction, permanent progress in Guadua processing and products has been achieved.

The special providers usually maintain stronger relationships to the meso-level actors than the SMEs, which result from the size of their enterprise, their formal registration and their pool of Guadua related knowledge and experiences. Most of these actors are integrated in institution building processes, and some of them also act as instructors and speakers during meso-level E&E activities (*Table 58*). Furthermore, the meso-level actors conduct on-farm and in-enterprise R&D, and the special providers contribute their expertise to meso-level R&D activities. In the period between 2002 and 2004 new special providers entered the Guadua sector. Three of them were investigated in detail, revealing that one evolved from a community Guadua project initiated at the meso-level. The other two were private enterprises, with one of them had close contacts to a research entity. The mutual interactions seemed to be of greater interest and benefit to the special providers and the meso-level actors than the common SMEs.

Conclusions

In the investigated SMEs the most important source of knowledge for the adoption of new products was the spill-over from the enterprises, which resulted from the technological knowledge commonly available to all actors. All actors contributed to this freely available knowledge pool with trial and error and modification activities (R&D), and the sale of products in the market. Co-operation with other entrepreneurs and meso-level actors was not identified as an important source of knowledge and innovation. However, all the entrepreneurs knew about the meso-level E&E activities offered and had participated at least once.

The institutions currently governing the Guadua sector were mainly perceived as either irrelevant to the SMEs or as promoting the SMEs. Only a few interviewees criticised these institutions, referring to the overregulation of the resource use and the Guadua product trade. The informal status of most SMEs in the sector may be a reason for the weak interaction between the micro and the meso-level. Larger SMEs exhibit other patterns, which are characterised more by institutional and inter-level interactions.

As the results of the meso-level innovation and SME promoting activities, as well as the impact of the institutions in place, were low, the driving force behind innovation was assumed to be the enterprises' individuality and limited knowledge, which define their interaction capacity, their level of information and their market access.

The large-scale providers revealed patterns of interaction and innovation different to those of the SMEs. These differences originated partly from the size of their businesses, which compelled regular interaction and formality, and their endowment with capital and market experience in the agricultural sector. The meso and the micro-levels interacted regularly, with both sides benefiting from this interaction.

9.2. The Colombian Bamboo Sector in the international Context

An analysis of the bamboo sector case studies conducted so far revealed a development gradient from east to west, from Asia to Latin America to Africa. Whereas in Asia the heterogeneity between the countries is great, e.g. between the Nepalese, the Indian and the Chinese bamboo sectors, in Latin America and in Africa the standard in the bamboo sectors of the various countries is similar within the respective continents. However, the Asian countries are much more advanced in relation to industrial uses of bamboo and the establishment of bamboo plantations. Generally, in these countries the legal and institutional arrangements governing the use of the bamboo resource are clearer and more advanced.

The bamboo sectors in the Latin American countries are in a phase of transition. The interest in this resource is increasing. However, there are still deficits in the legal and institutional framework and the willingness of private actors to invest. The bamboo sector competes with the agricultural sector and other alternative investment possibilities. In the African countries, the bamboo sectors lack institutional recognition and the users lack the appropriate knowledge and skills in relation to the sustainable management and use of the bamboo resources. The African approaches to the promotion of the utilisation and processing of the bamboo resources are mainly community-based, as raw materials are usually common or state property, not private. In *table 60*, the basic conclusions of the most important bamboo case studies are listed and compared. Due to its outstanding position, the Chinese bamboo sector is not included in this list. Details relating to this sector are described in *chapter 1.2*.

Table 61. Common conclusions and recommendations of national bamboo case studies

Conclusions	Colombia[656]	Costa Rica[657]	Asian case studies[658]	African case studies[659]
Problems related to the value-adding processes and the micro-level actors (entrepreneurs and farmers)	- Low economic recognition as product and raw material - Mainly low value products and applications - No industrial demand and substitution through other materials - Artisan and informal character of the manufacturing sectors - Low private investment and technological development	- High opportunity costs of small to medium size bamboo plantations and enterprises - Low price and low value products - No industrial demand - Artisan character of the sector - Informality of the sector - Disorganised actors	- Lack of resource management skills - Increased industrial and export demand threaten sustainability - Lack of entrepreneurial skills - Lack of private investment	- Lack of resource management skills - Lack of processing skills - Naturally not widely spread. Not widely known as resource for subsistence and commercial uses - Low level of industrial processing - Disorganised users - Production of low-value products

[656] As in HELD AND MANZANO (2004)

[657] As in DERAS (2004) and ADAMSON AND LÓPEZ (2000)

[658] As in SINGH, BERSALONA AND QUINTANS (2000); MAOGONG ET AL. (1998); KARKI, SHERCHAN AND KARKI (1998); SRIVASTAVA (1998); MATHEW (1998); and HANUMAPPA (1996). Without China. For a detailed description of the Chinese bamboo sector see *chapter 1.1*.

[659] As in ESEGU, SSENTEZA AND SEKATUBA (2000); KELBESSA ET AL. (2000); ONGUGO ET AL. (2000); and TAFORI (2000)

Table 61. Continued

Problems related to the institutional framework and the meso and macro-level actors	- No recognition as major forest product, thus no subsidies - Disorganised actors and many stakeholders -Mainly regulatory, fewer promoting institutions - No public investment	- Massive institutional promotion in the 1980s failed due to personal and financial mismanagement - Lack of institutional instruments for trade and production promotion	- Lack of resource management institutions - Fragmentation of the bamboo actors - Lack of SME promotion - Lack of public intervention	- No legal recognition as utilisable forest resource - Mainly protection regulations - Lack of public investment and R&D - Lack of SME promotion
Recommendations	- Develop products and processes for export and industrial use - Create incentive and subsidy structures for plantations - Promote export of bamboo products - Promote bamboo related FDIs and foreign R&D - Formalisation of the enterprises - Provide capital and funds through easy access credits and public programs - Create a bamboo productive chain/cluster approach	- Increase value of the raw material and the products - Promote industrial demand and the production of exportable products - Amplify the fields of applications for bamboo, also through scientific research - Promote and incentive formalisation of the enterprises - Promote investments - Organise the sector and formulate a plan of action - Technology transfer	- Extension and training in resource management and entrepreneurial skills - Technology transfer - Formation of a national bamboo sector approach - Diversified and value-added products - Integrate research entities and intensify research - Create incentives and subsidies - Provide credit schemes	- Extension, management training and on-farm training for communities and users - Improve processing technologies - Definition of bamboo resources in the institutional context - Establishment of appropriate formal institutions - Organisation of users - Encourage and promote community-based small scale artisan enterprises

All the case studies recommended strengthening the technological development and the improvement of processing and entrepreneurial skills. The case studies identified in the underdeveloped value-adding processes of the bamboo sectors' micro-level actors (farmers and entrepreneurs) as the major weakness hindering a greater contribution to rural development processes by these sectors. The basic analysis of the Colombian bamboo sector revealed numerous similarities to other bamboo sectors all over the world.

Basically, all of the case studies highlighted the entrepreneurs' lack of value-adding capacity, the problems of meeting a compromise between resource protection and resource utilisation and the missing institutional support for SMEs and bamboo product trade. In all of the case studies, the institutional framework and the related actors were called upon to promote the micro-level and to facilitate through adapted institutional arrangements the improved access to promotion schemes, capital, information and markets.

The promotion strategies recommended by the authors of the national bamboo case studies were based on experience made with the development of forest based sectors in Europe and North America. During their development the forest-based sectors in the industrialised countries have undergone organisational and production changes over a period of several hundred years, such as[660]:

- Early integration of specialised firms, i.e. of sawmills and the pulp industry, leading to inter-firm innovation networks (19th century)
- Upgrading of product quality, i.e. the quality of paper, leading to increased market shares and consumer preferences
- Co-operation with other industries, such as the chemical industry and the machinery industry, leading to inter-sector innovation networks and access to markets
- Aggressive human capital policies and knowledge network promotion by the government (20th century)
- Increased domestic as well as export demand

The promotion strategies of the meso-level actors in the Colombian bamboo sector draw mainly from these experiences, assuming that the state of development of the bamboo sector is similar to the early stages of the European forestry sectors. The major factors in the successful economic development of the European forestry and wood processing sectors were seen as their close co-operation with other industries. This co-operation not only promoted technological progress and improved the competitiveness of the forest product sectors, but the other sectors also profited, as e.g. the Scandinavian machinery industry, which today leads in the production and export of timber harvesting and pulp and paper machinery, or the chemical industry, which leads in paper chemical production[661]. MÄKINEN AND SELBY described the success factors in the Finnish forest cluster at the SME level. They concluded that direct marketing strategies and client co-operations are of particularly great importance to a firm's competitiveness, as well as its ability to create and offer new products and services[662].

According to PORTER'S competitiveness approach, the original availability of the resource base (in this case the forest resources) and the early decision of the firms to co-operate determined the success of the Scandinavian forestry sectors. The absence of advanced factor conditions did not seem to affect the sector's development negatively in this case. Nevertheless, it probably slowed down the rate of the forest sector's development, which took

[660] According to BONITA ET AL. (2002: 4) and FERRANTI ET AL. (2002: 6 and 53ff.)

[661] BONITA ET AL. (2002: 4)

[662] MÄKINEN AND SELBY (1996: 71ff.)

about 300 years from the 16th century to its peak in the mid 20th century. Today, other industries are central to the Scandinavian economies. However, most of these industries originally grew through their activities in the forest products sector[663].

The competitiveness of the German forest products sector was analysed by THOROE AND OLLMANN (1998). They found that in Germany, the forest products sector is facing increasing competition from the forest sectors of neighbouring European countries, and from Asia and North America. The international competitiveness of the German forest sector is closely related to the business location and the immobility of the most important production factors of these sectors: the resource base and labour. Consequently, the forest product sectors compete with the comparative advantages of their respective business location in attracting the production factors capital and knowledge. One major strategy of the German forest sector is the application of highly efficient technology in the production process and the innovative application of the raw materials[664].

In most Latin American forest sectors, these experiences have not been replicated as yet. The reasons for this can be found in the poor foundation for technological progress laid in the colonial period and the cultivation of inefficient forest manufacturing sectors in the post-war period (from 1945)[665]. During the colonisation and cultivation of the Latin American countries the extraction of natural resources and agricultural production were promoted primarily. The transformation of forest into arable agricultural production land is still preferred to investment in forest management and related value-adding processes. However, competitive wood processing sectors can be found in Brazil and Chile, for example, where the production of pulp and paper, the establishment of plantations and the furniture producing sectors have revealed itself to be a feasible economic investment opportunity, which is mainly based on the strong domestic markets and economies in these countries.

Innovation is of crucial importance to the value-adding process in the forest resource processing sector. In the management of the resource base itself, process innovations dominate, whereas in the processing sectors process and product innovation are equally important[666]. The experiences from the Scandinavian forest clusters have revealed the great importance of inter-actor and inter-sector links for successful technological progress and increased international competition. Several forestry and timber processing sectors of the Scandinavian, Central European and North American countries have been studied in relation to their contribution to the regional and national economies and countries, and their competitiveness as economic sectors by the EFI[667]. The result was that advanced value-adding processes are based on technological processes and innovative products[668]. Generally,

---

[663] FERRANTI ET AL. (2002: 6)

[664] THOROE AND OLLMANN (1998: 244)

[665] FERRANTI ET AL. (2002: 61f.)

[666] SCHMITHÜSEN ET AL. (2003: 127)

[667] See HYTTINEN, MONONEN AND PELLI (1996)

[668] In contrast to the forestry and wood processing sectors in developing countries as described in *chapter 6.4.3*.

the wood processing firms were found to be much more active in specialisation and product innovation than the raw material producing forestry sectors. A study of medium and large-scale German forestry enterprises revealed that these types of enterprises mainly adopt process and organisational innovations[669].

In comparing the current state of the Colombian bamboo sector with the experiences made in the forestry and wood processing sectors of the industrialised countries, the main shortcomings were identified as the lack of interaction between the bamboo-based enterprises and other industry sectors and branches, the absence of the horizontal integration of bamboo-based enterprises and a market, which is mainly demanding low value bamboo products. The results of this study showed that the enterprises themselves are continuously working to improve product quality, product range and production technology. However, the interaction and co-operation with other industries, other bamboo based enterprises and clients might have speeded up this development process.

The enterprises in the Colombian bamboo sector are mainly informal artisan enterprises without common organisation. Their endowment with physical and human capital is low and their business options are limited. During the development of the European forestry and timber processing sectors, the national industries co-evolved and the forest-based enterprises were at least on an equal footing with enterprises from other industries. The inter-industry interactions and co-operation were characterised by mutual economic benefits. Today's Colombian bamboo sector finds itself in an inferior position compared to the Colombian chemistry, oil or metal industries and the powerful agricultural production chains. Hence, inter-industry co-operation is linked to investment flows from the advanced industries to the bamboo enterprises. This option is currently not perceived as a feasible business opportunity by the other Colombian industrial sectors.

The gap between the Colombian industrial sectors in terms of enterprise structure and capital endowment is a hindrance to the interactions, which have been successful in the case of forest sector development in other countries, where the level of economic capacities between the enterprises of the sector have been more equal. The aims of the meso-level actors should be 1) to promote the bamboo sector enterprises to advance further in terms of technology and organisation and to approach the other industry sectors, and 2) to encourage the advanced industries to invest in co-operation with bamboo based enterprises.

---

669 PICKENPACK (2004: 179)

9.3 The Colombian Bamboo Sector in the national Context

URREA ET AL. (2000) compared the different patterns of entrepreneurial organisation and innovation in Colombia's industries. They distinguished between three Colombian regions, paying tribute to the assumption that the geographical, cultural and historical features of a region influence and shape the entrepreneurial behavioural patterns. In the context of this study, the Colombian bamboo sector may be compared to other economic sectors located in the Colombian Andean region[670]. The study by URREA ET AL. was based on DANE and COLCIENCIAS statistics and publications relating to the analysis of economic sectors under considerations of innovation research and upgrading processes.

The industrial sectors in the Andean region are basically constituted of SMEs and micro-enterprises. The figures provided by URREA ET AL. revealed high numbers of firm mortality amongst the micro enterprises and SMEs; more than 60% of enterprises were founded less than three years ago. These SMEs employed the most labour and were responsible for the highest value-adding in the region. Between 1993 and 1996, 75% of the enterprises undertook innovation activities. The innovation management of these enterprises was based upon three factors: 1) the adoption of externally developed products and processes was favoured over investment in entrepreneurial R&D activities (only 25% of the enterprises conducted R&D); 2) incremental innovations were favoured over radical innovations (mainly product quality and design improvements); and 3) the capital invested in the innovation processes was firm capital (only 18% received credit for purposes related to innovation activities). In general, the acquisition of foreign technology and the re-sale of innovative foreign products in the domestic market was favoured over engagement in R&D activities and the marketing of domestic innovations (40% of the product innovations entering the Colombian market annually originated from Colombian enterprises).

Compared to the situation described for the industries in the region, the Colombian bamboo sector revealed similar patterns in terms of the composition of the sector constituting enterprises. The sectors are mainly constituted of micro enterprises and SMEs. Their innovations are mainly incremental and access to external sources of capital for investments is limited. However, the SMEs constitute the backbone of the sectors and provide more employment possibilities than the large-scale enterprises. The major differences between the bamboo and the other industrial sectors in the Colombian Andean region are: 1) the small total number of enterprises (SMEs and large-scale) active in the bamboo sector; 2) the geographic factors restricting the bamboo sector to the natural resource base; and 3) the low level of the common technological knowledge available to the enterprises in the bamboo sector, which is restricted to artisan production processes and low value applications in the construction sector.

---

670 URREA ET AL. (2000: 221ff.)

The development goal articulated by the meso-level actors for the bamboo sector is its industrialisation organised in the form of an agro-industrial production chain. The innovation management of these agro-industrial sectors usually comprises a collaborative or state research entity, which conducts R&D in the various fields important for production, processing and marketing of the respective agricultural production chain. The individual producers and processors of these sectors are deposed from their SCHUMPETERIAN 'duty' to conduct R&D and to innovate. However, the bamboo sector is not yet organised in such a production chain, nor is it endowed with a collaborative research entity.

URREA ET AL. recognised the socio-cultural influence on the innovation activities of the entrepreneurs, such as the employment structures and whether employees may become partners or self-reliant entrepreneurs, which is essential for innovation and knowledge diffusion processes in a sector or a region. URREA ET AL. saw this as a factor of increasing importance in innovation processes. However, the basic and professional education in the enterprises continues to be deficient and hinders this diffusion path. Another socio-cultural aspect is the traditionally weak relationship between research entities (universities) and the manufacturing sectors in Colombia[671]. In the study of the Colombian bamboo sector these cultural patterns could be detected too. The weak links between the private SMEs and the universities and research entities, in particular, were confirmed. The importance of fluctuating employees for knowledge and innovation diffusion, on the other hand, was not observed.

With respect to the sources of innovation and innovative ideas in the industrial enterprises, URREA ET AL. identified both, sources in the enterprises themselves and external sources. The most important sources were international exhibitions and seminars, part of the spill-overs, followed by the extension and dissemination activities of meso-level actors. Interaction and co-operation with research entities and universities was revealed to be of little importance for innovation. These interactions are weak, and are mainly maintained by large enterprises rather than by SMEs. This observation was also made in the bamboo sector, where the large-scale producers are integrated into R&D activities and institution building processes, whereas the SMEs are perceived as clients of extension and education events. Again, there are clear similarities between the bamboo-based SMEs and SMEs from other industries. However, the industrial SMEs seem to benefit more from E&E activities than bamboo SMEs. This may have to do with the level of knowledge disseminated, which is not very advanced in the bamboo sector.

The low degree of inter-firm and inter-sector interaction not only affects the bamboo-based SMEs, but all SMEs in the Colombian Andean region. However, the small size of the bamboo sector (in terms of the total number of enterprises and regional extension), its deficient value-adding processes and products, and its weak outside links isolates the bamboo sector more than the enterprises from other sectors. The poorly advanced technology in the bamboo sector

[671] URREA ET AL. (2000: 263ff.)

also enhances the isolation effect. Therefore, the current low technological level of the bamboo sector is not only the result of the absence of interaction and co-operation, but also the result of the weak structural condition of the sector itself, which negatively affects the technology and capital exchange with other industries and firms. Generally, the larger firms co-operate more actively with meso-level actors, entrepreneurs and other industrial sectors than the SMEs. The meso-level actors should consider the structure and condition of the entrepreneurs and enterprises in the bamboo sector when implementing their promotion strategies as these enterprises are at least partly incapable of fulfilling the expectations placed upon the bamboo sector.

9.4 Promotion of entrepreneurial Innovations

The Colombian bamboo sector has proven to be a 'role model' survival cluster, as defined by ALTENBURG AND MEYER-STAMER (1999). It is characterised by micro-enterprises and SMEs, the informal status of these actors, their limited knowledge (caused by information asymmetries and a low degree of technical and entrepreneurial expertise), low-value adding processes, and a general deficit in capital and investments.

Technological progress has been explained in theory as the driving force behind economic development. In the neo-classical and in the new growth theory the origins of technological progress could only be partly explained and have been mainly reduced to the R&D activities, R&D related expenses and R&D employment figures of both, public and private actors. The research results of this study show that in the bamboo sector R&D is not the most important source of innovation, but the passive interactions between the micro-level actors in the form of spill-overs. The entrepreneurs in the investigated sector are limited in their access to input factors required for R&D, such as capital, knowledge and labour. Therefore, the meso-level actors have adopted R&D and R&D promotion into their sector promotion strategies.

The meso-level actors of the Colombian sector see in the intensification of R&D activities the main means of promoting the sector's SMEs. They have placed an emphasis on the development of industrialisation technologies and products competitive in the international markets. Although such technologies and products already exist, the issue of technology and knowledge transfer from other producing countries and regions has only been considered to date, but not yet achieved. This is mainly due to the enterprise structure encountered in the bamboo sector and their current state of technological, as well as economic expertise, which have not yet attracted Colombian and foreign investors. The attraction of such investors has not been integrated into the meso-level sector promotion strategies, as they assume this to be the responsibility of macro-economic politics. Nevertheless, through the promotion of the business location 'Bamboo sector' in terms of intensified R&D and formal organisation they seek to provide attractive conditions for such investors. As the focus of this study has been on the promotion of SMEs and their innovation capacity, this aspect has not been exhaustively investigated.

The integration of the SMEs in collaborative R&D projects is not common as a means of promoting R&D in the bamboo sector's micro-level, although this aspect is perceived as crucial to entrepreneurial innovation research. The co-operative on-farm or in-enterprise R&D is rare. The large-scale bamboo producers are integrated into such forms of co-operation, however. The integration of the large-scale producers is perceived as mutually beneficial as these producers have the capital and advanced knowledge necessary to contribute significantly to collaborative R&D. BOUDOURIDES (2001) described such large-scale producers as central actors, who function as multipliers of advanced technologies and products, and as role models for other entrepreneurs within a sector. However, the gap between the sector's SMEs and the large-scale producers in terms of capital endowment and products cannot easily be bridged by the SMEs. Currently, the large-scale producers and meso-level actors are expanding their R&D activities, whereas the SMEs cannot follow due to lack of resources and their business strutures.

The SMEs are mainly offered extension and education activities and promotion measures, which enhance entrepreneurial R&D, such as credit schemes and promotion networks. The research has shown that these measures are not suitable for bamboo enterprises, since these instruments assume the formal status of the applicants. Furthermore, the respective meso-level actors do not perceive the bamboo-based business ideas as promising and feasible. In total, the bamboo based sector is perceived as low-value, not innovative, informal and as a subsistence sector.

The majority of the sector's SMEs draw upon the commonly available knowledge pool in the bamboo sector, which is restricted to artisan manufacturing techniques and basic applications in the construction sector. The research results show that the entrepreneurs are modifying this common knowledge with their individual tacit knowledge, resulting in incremental product and process innovations. This mixed knowledge becomes available to the other entrepreneurs when the new products enter the market and can be purchased or copied (MCKELVEY 1998). Hence, the technological knowledge pool of the bamboo sector is growing slowly. On the other hand, large numbers of entrepreneurs are leaving the bamboo sector after only a short period of time and take with them the knowledge they have acquired. The knowledge pool of the SMEs is not advancing at a rate matching the development goals of the meso-level.

The research has shown the investigated SMEs tend to stick to an innovation path (compare DOSI, PAVITT AND SOETE, 1990), which is optimal based on their knowledge. The meso-level development goal is leapfrogging several intermediate stages in this innovation path. With the promotion activities offered and within their structures, the SMEs cannot achieve the development goal defined by the meso-level. Furthermore, the SMEs are in danger of being bypassed by the sector development and promotion strategies. COOPER (1980) identified these discrepancies also, while investigating innovation in agricultural sectors in developing countries.

In Europe and North America, the agglomerations of related firms and their close social and economic interactions in the form of co-operation and competition have been identified as important driving forces behind the competitiveness of the individual firm and the respective economic sector as a whole. The mix of competition and co-operation was deemed crucial to product, process and organisational innovation processes and the most effective way to diffuse the knowledge and experiences made through R&D within an economic sector. The economic policies to support and promote economic structures and growth have adopted these experiences, wherein the promotion of the interaction among entrepreneurs, among institutional actors and between these two actor groups in form of regional and national innovation systems and networks holds an essential position. In the Colombian bamboo sector the promotion and enhancement of the interactions and the organisation of the sector's actors is still in the initial phase. The institution building and organisation processes require resources, which in the long run only the large-scale producers are willing to invest. On the other hand, they depend on and benefit most from these processes and have a great interest in influencing them.

There is the notion amongst the micro-level actors that the institutions currently planned will result in an over-regulation of the sector and hinder free entrepreneurial development, including their freedom to innovate and adopt products and technologies. In a sector, where the production processes and products vary greatly, as is the case in the Colombian bamboo sector, it is doubtful whether such institutions will make the sector more competitive and coherent, or whether an improved SME promotion strategy would be more suitable. In the current situation the bamboo sector's micro-level, the development progress and enterprise structure of which is characterised by features typical of the economic sectors of developing countries, is facing a meso-level constituted of a large number of fragmented actors, who are only partly co-operating and co-ordinating their activities. Furthermore, governmental actors, whose interaction patterns are hierarchical with respect to the micro-level and the other meso-level actors, dominate the meso-level. Basically, the bamboo sector's meso-level exhibits features described by MEYER-STAMER (2001) for meso-levels in industrialised countries.

In the course of the theoretical work concerning innovation in forest resource-based sectors in developing countries, no studies or research works could be found, which focus on the analysis of sector promotion strategies to enhance the innovation capacity of the entrepreneurs in these sectors. Commonly, the diagnosis of value-adding processes is picked as the central theme of such works. Such value-adding processes, which are usually diagnosed as deficient, are seen in relation to deficits of the entrepreneurs in terms of access to information and technical and entrepreneurial skills. Hence, the extension and dissemination of these items is recommended in the sector promotion strategies. Additionally, micro-credit schemes are implemented, which enable the entrepreneurs to acquire machinery and employ personnel.

As mentioned before, access to credit schemes and support programmes is limited for Colombian bamboo based SMEs. Furthermore, the integration of these enterprises into R&D is low. Hence, the extension and education activities of meso-level activities are perceived as the most import promotion instrument for the bamboo sector. As the research results have shown, the SMEs actively draw on these activities. However, there is a two class society of entrepreneurs, the first are willing and/or able to pay for such E&E measures, whereas the second are not or cannot. The E&E activities could not be confirmed as important in terms of knowledge diffusion and the initialisation of entrepreneurial R&D, trial and error, and innovation processes. The innovation relevant inputs mainly originate from inter-entrepreneur spill-over effects. The E&E activities are important in the dissemination of standard knowledge, which is still not very advanced. Naturally, there are exceptions and there are very advanced courses and workshops, usually offered to architects, civil engineers and constructors by the universities. The majority is perceived as receivers or paying clients. Only a few, again experienced large-scale producers and processors, are integrated into E&E activities during field trips or as instructors in courses.

In conclusion to this discussion chapter, there remains the question of which micro-level actors can achieve the development goals defined for the bamboo sector by the meso-level actors. The SMEs, in their current structure and state of development, require intensive promotion measures and organisational changes to achieve these goals. External (domestic and foreign) investors and entrepreneurs are an option, but have not yet been approached by the meso-level actors. Apparently, the potential economic profitability of the bamboo sector's business options alone has not yet convinced entrepreneurs to investment in the sector. Currently, the large-scale producers are the technically most advanced entrepreneurs in the bamboo sector and their greater integration into sector promotion activities benefits the sector as a whole. The bypassing of the majority of SMEs, on the other hand, limits the number of potential adopters of new technologies and products, and is slows the development process in the sector. Finally, discussing the results of the present research, one should consider the conclusion drawn by VERBECK (1999; Compare *chapter 3.2*) from his study on the promotion of innovations in European industrial clusters that innovative behaviour in general is much more resistant to promotion activities and than it has been assumed so far.

9.5 Effects of intra and inter-level Interactions on the Promotion of SMEs

According to the theoretical approach of systemic competitiveness, and other models of cluster and economic sectors, the co-ordination and co-operation of the meso-level and micro-level actors is assumed as a basic asset of successful sector development processes. However, according to the actor-centred analytical research approach, these interactions underlie a perceived reality, which the actors aim to shape and optimise according to their needs and potentials, but which also determines their room for action.

Actors of the meso-level interact along a gradient between co-operation and competition. Both options enable the meso-level actors to improve their endowment with resources. Co-operation is aimed at the optimisation of the employment of the actors' resources. The goal is to enhance the effectiveness of the actors' activities through synergistic co-operation, which is achieved by the optimal combination of each actor's expertise and access to technical, financial and institutional resources. Therefore, co-operation is the avoidance of parallel lines of work and enhances the diffusion of knowledge. The option to compete is the attempt by the actor to increase his access to resources, which is generally linked to the loss of resources by another actor in the same field of interest. Competition can foster the improved quality of the actor's work and activities. On the other hand, the actors are dependent upon the resource donors (public funds or private clients), which may influence the direction of research, contents of extension activities or the basic thematic orientation.

The Colombian bamboo sector's meso-level actors practice a mix of competing and co-operating strategies. In the field of R&D, the willingness to co-operate is high, since the basic level of common knowledge is low and through co-operation the perceived mutual benefit of knowledge acquisition is high. However, once a certain level of individual knowledge is reached, and is perceived as 'tradable' product, R&D in the respective field tends to be conducted individually and interest in co-operation decreases. Examples of this are the development of machinery and the development of bamboo propagation techniques. The extension and dissemination of R&D results and common knowledge is closely linked to the individual actor's R&D interaction strategy. The meso-level actors co-operate to conduct these E&E activities when the knowledge to be disseminated is general. When a higher level of individual knowledge is reached, the E&E activities become lucrative sources of income, especially for the research entities, who in turn seek to cover the investments they have made in garnering this knowledge. The meso-level actors mainly conduct these extension activities individually.

With respect to the I&I-activities, the meso-level actors depend on the guiding and facilitating role of governmental actors, who in turn compete for their outstanding position in the bamboo sector. As the micro-level actors are still not organised in a production chain, the governmental actors have not yet given control over resources to them. Furthermore, the intermediate position of bamboo between natural resource and private agricultural production makes the sector a playground for a comparably large number of governmental actors claiming responsibility for the resource and the related institutional framework. Due to this situation, the level of interaction in terms of institutions-building is high. However, it is characterised by the dominance of governmental actors and by competition between the meso-level actors for the limited resources to be distributed for bamboo sector related purposes by the state and international donors. The reduction of the number of meso-level actors involved might ensure more effective sector promotion and institutions-building. This number, however, should be above the critical mass to maintain the development and promotion processes, but below the current number, which includes actors, who are not endowed with the necessary expertise and resources.

The R&D related sector promotion strategies are aimed mainly at the dissemination of R&D results. The active integration of micro-level actors through on-farm and in-enterprise R&D

co-operation is only found with large-scale producers. This type of R&D co-operation is perceived as more mutually beneficial than co-operation with SMEs or micro-enterprises. Only in micro-projects is R&D conducted jointly by the meso and micro-level in form of the application of newly developed machinery and propagation techniques. Generally, the micro-level is perceived as a receiver of R&D results, rather than as co-operation partner. This is similar in E&E, although occasionally experienced entrepreneurs are invited as guest speakers or instructors.

The micro-level actors mainly compete with each other in the market. However, these interactions are also an important source of innovation and knowledge. Spill-over is the major form of diffusion in the sector. Nevertheless, early stages of co-operation also occur, such as the mutual lending of machinery and materials, as well as consultation in professional matters. Close co-operation between SMEs and suppliers or clients could not be identified, although certain SMEs have a stock of permanent clients, whose demands influence the design and quality of the bamboo products. The large-scale producers combine raw material production and various steps of processing within one enterprise and demonstrate the advantages of such integrated production and processing. Furthermore, they maintain close client relations, from which they draw demand driven impulses for the modification and adoption of new products and processes.

The interaction between the micro and the meso-level is top-down rather than interactive, which is enhanced by the lack of organisation or representation of the micro-level's enterprises. The competitiveness of large-scale producers benefits more from meso-micro-level interactions than that of the SMEs, which is mainly due to the different organisational and managerial structures of the two enterprise types. Basically, the meso-level interaction efforts reach the micro-level's SMEs. The effect these interactions have on the SMEs' competitiveness, however, is limited. The contents of most of the available education and extension activities, which can be afforded by the majority of the entrepreneurs, are not very advanced compared to the common knowledge in the bamboo sector. Extension offers beyond this level are scarce and mainly involve participation fees. The meso-level's E&E activities should widen the perspectives of the participants with regard to the amplification of technical and entrepreneurial knowledge, knowledge beyond the boundaries of the bamboo producing and manufacturing sector, e.g. machinery, electricity, wood product markets, etc.

At the micro-level, interaction is only weakly developed and mainly restricted to passive spill-overs and an irregular exchange of professional experience. Compared to enterprises investigated in European cluster studies, the SMEs in the Colombian bamboo sector are not in the sector long enough a) to develop strategic horizontal co-operation with other producers or vertical co-operation with suppliers or clients, and b) to advance significantly in terms of knowledge and technology. The enterprises working in the sector longer exhibit organisational and managerial features with only limited potential to achieve the meso-level's development goals for the sector. Thus, the meso-level actors have difficulties finding SMEs in the sector with which they can establish mutually beneficial co-operation. The empirical research has shown that a large number of meso-level actors are active in the bamboo sector.

These actors, however, have little contact to actors from other industrial branches or sectors. Hence, inputs and experience from other more advanced Colombian sectors, such as the chemistry and machinery industry, are not transferred to the bamboo sector. Weak interactions are maintained with the Colombian forestry and wood processing sector, which itself is poorly organised and not very competitive.

9.6 Concluding Remarks

The analytical approach to the investigation of the Colombian bamboo sector according to ESSER ET AL.'s concept of systemic competitiveness has basically proven to be useful. The main research questions were answered. However, the research's restriction to the meso and the micro-level limits the interpretability of the research results. General statements on the innovation behaviour of the micro-level actors and the promotion strategies of the meso-level actors must take into consideration macro-political and economic framework conditions. These conditions were presented and investigated in this study work, but their direct influence on the research objects was not empirically investigated. The same restriction must be considered in relation to information about the meta-level influences on the micro and meso-level actors. The meta-level's influence was qualitatively assessed, but could not be analytically incorporated into the research. Thus, its main use was to amplify the interpretation context of the research result.

For the purposes of this study the analysis of the two levels, micro and meso, provided adequate results for theory and practice. For research with a greater emphasis on interaction research than on innovation research, a detailed analysis of the meta-level is recommended. For research emphasising entrepreneurial development, the macro-level conditions and its influences on entrepreneurial decision-making should be considered.

This research placed an emphasis on the analysis of the meso-level using the actor-theoretical approach of actor-centred institutionalism by SCHARPF and MAYNTZ. The strengths of this approach have been in providing explanations for the driving forces underlying the intra-meso-level interactions, i.e. in explaining the resource-oriented interaction strategies of the actors and how these influence sector promotion strategies. Restrictions exist in relation to the explanation of interactions between individual micro and complex meso-level actors. The research results achieved using the actor-centred approach could potentially gain in validity and explanatory power if real resource distribution streams (monetary and institutional resources) were to be surveyed. This could underline the qualitative explanation of the patterns of interaction between the meso-level actors.

In the investigated sector, the SMEs and micro-entrepreneurs behave as actors influenced by their socio-economic environment rather than by the market influence, whereas the decisions of large-scale entrepreneurs are integrated into and influenced by the market. Classical innovation research focusing on R&D oriented business and innovation management is a useful tool in the context of these large-scale entrepreneurs/special providers of Guadua products. The evolutionary innovation research approach is useful for the interpretation of the innovation behaviour of SMEs. In the case of these actors, profit optimisation is not the decisive element of entrepreneurial decision making, but the mix of their own tacit knowledge and common knowledge, which is limited by the lack of capital and the market situation.

The innovation process in the enterprises is initiated by a mix of interactions through which ideas and knowledge are acquired, complemented by the available pool of common knowledge and technology. Innovation and diffusion processes take place along paths, which are determined by the restricted knowledge of the enterprises. Hence, innovations promoting activities have to consider these paths and not bypass or overhaul them placing a strain on the micro-level. Interactions between the micro-level actors are mainly restricted to spill-over and low-level co-operation. The diffusion of knowledge and innovation within the sector follows individual adoption strategies. The theoretical advantages of firm agglomerations resulting in an optimal mix of competition and co-operation between them is only limitedly transferable to the sector investigated.

For further research in the field of innovation and sector promotion in economic sectors in developing countries, the following research foci are recommended to round off the results of this study:

- An in-depth investigation of the influence of meta-level factors on the interactions between meso-level actors, between micro-level actors and between the two levels, i.e. considering the aspects of trust and conflict.

- Regular innovation research surveys in forest resource-based sectors in developing countries according to codified criteria and indicators, as for example, fixed by the OECD's Oslo Manual and modified according to the conditions encountered in developing countries.

- An investigation of the micro and macro-economic impact of the informal status of enterprises and how institutions can formalise these enterprises without depriving them of their economic foundations and in turn destroying employment structures.

- An investigation of the meso-level actors applying network and organisational theories to analyse the internal structures of these actors and the influence these internal structures have on the actors' strategic orientation and sector promotion activities.

- An investigation of the survival clusters of developing countries to identify more detailed socio-economic features distinguishing them from other agglomerations of economic units, which in turn may help to initiate more adequate promotion instruments and programs.

- A study of the paths of diffusion within forest resource-based sectors and survival clusters to reconstruct the key interactions for diffusion and develop promotion measures, which ideally enhance these key interactions.

- An investigation of the reasons for firm closures, especially in the case of 'successful' enterprises in order to promote their continuation in the sector.

- An investigation of the role foreign investors and enterprises play in the innovation processes of economic sectors in developing countries to draw conclusions for the promotion of such interactions.

- An investigation into the mix of R&D, E&E and I&I-activities, which are most successful in terms of promoting innovation at the micro-level and at the same time are most economic with respect to the meso-level resources.

10. Bibliography

ADAMSON, M AND LÓPEZ, D. (2000): Socioeconomic study for the bamboo sector in Costa Rica. INBAR Working Paper Series No. 32, pp. 120

ADKOLI, N. (2002): Study of the use of bamboo in the sericulture sector of South India. INBAR Working Paper Series No. 38, pp. 84

AKUMSI, A. (2003): Community participation in wildlife management. The Mount Cameron experience. Unasylva, Vol. 54, 214/215, pp. 37-43

ALTENBURG, T. AND MEYER-STAMER, J. (1999): How to promote clusters. Policy experiences from Latin America. World Development Vol. 27, No. 9, pp. 1693-1713

ALTENBURG, T., HILLEBRAND, W. AND MEYER-STAMER, J. (1998): Building Systemic Competitiveness. Concept and case studies from Mexico, Brazil, Paraguay, Korea and Thailand. DIE, Berlin, pp. 75

AMSDEN, A. (1989): Asia's next gain. South Korea and late industrialization. Oxford University Press, pp. 379

ANDERSEN, E. (1996a): Theories of localised resource-based growth and development. From Marshall to New Evolutionary Economics. In: HYTTINEN, P., MONONEN, A. AND PELLI, P. (eds.) (1996): Regional development based on forest resources. Theories and practices. EFI Proceedings No. 9, pp. 5-16

ANDERSEN, E. (1996b): Evolutionary economics. Post-SCHUMPETERian contributions. Pinter Publications London, pp. 238

ARBELAÉZ, A. (2001): Investigaciones sobre Guadua angustifolia Kunth realizadas en Colombia (1950-2000). Facultad de ciencias agropecuarias. Universidad Nacional de Colombia, Sede Medellin, pp. 92

ARROYAVE, M., MESA, P. AND LÓPEZ, L. (2000): Recuperación del conocimiento tradicional en el aprovechamiento de la Guadua en el municipio de Pereira. Convenio UTP-GTZ. Facultad de Ciencias ambientales. Universidad Tecnologica de Pereira, pp. 99

ATTESLANDER, P. (1995): Methoden der empirischen Sozialforschung. Sammlung Göschen. De Gruyter, pp. 418

BATHELT, H. AND GLÜCKLER, J. (2002): Wirtschaftsgeografie. UTB, pp. 319

BECATTINI, G. (1990): The Marshallian industrial district as a socio-economic notion. In: PYKE, F., BECATTINI, G. AND SENGENBERGER, W. (eds.) (1990): Industrial districts and inter-firm co-operation in Italy. International Institute for Labour Studies , Geneva, pp. 37-51

BEINTEMA, N., ROMANO, L. AND PARDEY, P. (2000): I&D agropecuario en Colombia. Política. Inversiones y perfil institucional. International Food Policy Research Institute, Washington, pp. 75

BELCHER, B. (1995): The role of bamboo in development. In: INBAR (1995): Bamboo, people and environment. Vol. 4: Socio-economics and culture, pp. 2-9

BELCHER, B. (1998): A production to consumption systems approach: Lessons from the bamboo and rattan sector in Asia. In: Wollenberg, E. and Ingles, A. (eds.): Incomes from the forest. Methods for the development and conservation of forest products for local communities, pp. 57-84

BELL, L. (1995): The impact of minimum wages in Mexico and Colombia. The World Bank Policy Research Paper No. 1514, pp. 48

BELTRÁN, P. AND QUIMBAYA, H. (1987): Comercialización y uso de la Guadua (*Bambusa Guadua* Humboldt et Bonpland) y estimación de danos causados por *Podischnus agenor* (Col. Scarabaeidae) en el Valle. Tesis de grado. Facultad de Ciencias agropecuarias. Universidad Nacional de Colombia, sede Palmira, pp. 179+Annex

BELTRÁN, P. AND PASTRANA, R. (1991): Estudio técnico, económico y de mercados del montaje de una plantaproductora de pulpa que utilice como materia prima la Guadua. Tesis de grado. Facultad de Ingenieria. Universidad Javeriana, Bogotá, pp. 102

BENNINGHAUS, H. (1994): Einführung in die sozialwissenschaftliche Datenanalyse. Oldenbourg Verlag, pp. 464

BENTLEY, J. AND BAKER, P. (2000): The Colombian coffee growers\` federation: Organised, successful smallholder farmers for 70 years. odi agricultural research & extension network. Network paper No. 100, pp. 9

BERGMANN, E. AND FRESER, E. (1999): Industrial and regional clusters. Concepts and comparative applications. In: The web-book of Regional Science (www.rri.wvu.edu/reg-scweb.htm)

BERNAL, L. (2002): Estudio para definir criterios tecnicos y metodologicos para el inventario, ordenamiento, zonificación y manejo sostenibile del recurso Guadua en el Eje Cafetero, Tolima y el Valle de Cauca. Proyecto MSBC (Manejo Sostenible de Bosques en Colombia), Bogotá, pp. 85

BIANCHI, G. (1998): Requiem for a third Italy? Rise and fall of a too successful concept. In: Entrepreneurship and Regional Development, Vol. 10, No. 2, pp. 93-116

BIGGS, S. AND SMITH, S. (2003): A paradox of learning in project cycle. Management and the role of organizational culture. World Development, Vol. 31, No. 10, pp. 1743-1757

BOECKH, A. (1993): Entwicklungstheorien. Eine Rückschau. In: NOHLEN, D. AND NUSCHELER, F. (eds.) (1993): Handbuch der dritten Welt 1. Grundprobleme, Theorien, Strategien. Verlag J.H.W. Dietz Nachf., pp. 110-130

BOECKH, A. (1999): Wie man Unpassendes passend macht. Das Elend des Fortschritts in Lateinamerika. In: THIEL, R. (ed.): Neue Ansätze zur Entwicklungstheorie. DSE Themendienst 10. ZD, Bonn, pp. 82-95

BOHNET, M. (1999): Überlegung zur Zukunft der Entwicklungspolitik. In: THIEL, R. (ed.): Neue Ansätze zur Entwicklungstheorie. DSE Themendienst 10. ZD, Bonn, pp. 96-109

BOLLMANN, P. (1990): Technischer Fortschritt und wirtschaftlicher Wandel. Eine Gegenüberstellung neo-klassischer und evolutorischer Innovationsforschung. Physica Verlag, pp. 184

BONITA, M. ET AL. (2002): Forest clusters. A competitive model for Latin Aamerica. Inter-American Development Bank. Washington, pp. 63

BORTAGARAY, I. AND TIFFIN, S. (2000): Innovation clusters in Latin America. Paper presented at the 4th international conference on technology and innovation, held in Curitiba, Brazil from 28.08.-31.08.2000, pp. 40

BOSCHMA, R. AND LAMBOOY, J. (1999): Evolutionary economics and economic geography. Journal of Evolutionary Economics, Vol. 9, pp. 411-429

BOUDOURIDES, M. (2001): The politics of technological innovations. Network approaches. Conference proceedings of the International Summer Academy on Technological Studies: User Involvement in Technological Innovation, held in Deutschlandsberg from 8th to 13th of July 2001, pp. 31-41

BRENNER, T. (2000): The evolution of localised industrial clusters: Identifying the processes of self-organization. Max Planck Institute Jena. Papers for Economics & Evolution No. 11, pp. 22

BRENNER, T. (2002): Innovation and cooperation during the emergence of local industrial clusters – An empirical study in Germany. Max Planck Institute Jena. Papers for Economics & Evolution No. 205, pp. 19

BRENNER, T. (2003): Policy measures to support the emergence of localised industrial clusters. In: FORNAHL, D. AND BRENNER, T. (2003): The influence of co-operations, networks and institutions on regional innovation systems. Cheltenham: Edgar Elgar Verlag, pp. 325-350

BRENNER, T. AND GREIF, S. (2003): The dependence of innovativeness on the local firm population – An empirical study of Germany. Max Planck Institute Jena. Papers for Economics & Evolution No. 306, pp. 19

BRUSCO, S. (1990): The idea of the industrial district. Its genesis. In: In: PYKE, F., BECATTINI, G. AND SENGENBERGER, W. (eds.) (1990): Industrial districts and inter-firm co-operation in Italy. International Institute for Labour Studies , Geneva, pp.10-19

BUCHELLY, A. (2001): Sistematización del conocimiento y la experiencia en las áreas de producción y transformación de la Guadua en Caldas. Tesis de grado. Universidad Nacional de Colombia, sede Manizales, pp. xx

BÜHL, A. AND ZÖFEL, P. (2000): SPSS Version 10. Einführung in die moderne Datenanalyse unter Windows. Addison-Wesley Verlag, pp. 734

BUITELAAR, R. (2002): Aglomeraciones mineras y desarrollo local en America Latina. Alfaomega/CEPAL/IDRC, pp. 340

BUSSOLO, M AND LAY, J. (2003): Globalisation and poverty changes in Colombia. OECD Development Centre Working Paper No. 226, pp. 50

CAMARA DE COMERCIO DE BOGOTA (2002): Crear EMPRESA. Guia para construir y formalizar una empresa. Camara de Comercio de Bogota, Departamento de Desarroloo empresarial, pp. 98

CAMARGO, J. (in prep.)

Carlowitz, von H.C. (1713): Sylvicultura oeconomica oder Anweisung zur wilden Baumzucht. Leipzig, pp. xx

Carlsson, B. and Eliasson, G. (2001): Industrial dynamics and endogenous growth. Publication by the Royal Institute of Technology (KTH), Stockholm, Sweden, pp. 25

Carrizosa, J. (1992): La politica ambiental en Colombia. Desarrollo sostenible y democratización. CEREC, Colombia, pp. 220

Castaño, F. and Moreno, R. (2004): Guadua para todos. Cultivo y aprovechamiento. Proyecto MSBC (Manejo Sostenible de Bosques en Colombia), pp. 188

Cegli, G. and Dini, M. (1999): SME cluster and network development in developing countries. The experienceof the UNIDO. Paper presented at the international conference on: Building a modern and effective development service industry for small enterprises, held in March 1999 in Rio de Janeiro, Brazil, pp. 26

CEPAL (1990): Transformación productiva con equitad. CEPAL, Santiago de Chile, pp. 185

Chacón, J. and Zuluaga, J. (1990): Planeación de una pequeña empresa para la transformación de la Guadua en soluciones de vivienda para estratos socials bajos para el Valle geográfico del Cauca. Tesis de grado. Facultad de Ingenieria. Universidad Javeriana, Bogotá, pp. 125

Chica, R. (1996) : El crecimiento de la productividad en Colombia. Resumen de los resultados del estudio nacional sobre determinantes del crecimiento de la productividad. COLCIENCIAS, Colombia, pp. 30

CINEP (1998): Colombia. País de regiones. Vol. 1. CINEP/COLCIENCIAS, pp. 312

Clarke, G. et al. (2002): Bank lending to small business in Latin America. Does origin matter? The World Bank Policy Research Working Paper, No. 2760, pp. 36

Coase, R. (1937): The nature of the firm. Economica, Vol. 16, pp. 386-405

Colciencias (2002): Información básica sobre financiación de proyectos de investigación, innovación y desarrollo tecnológico en los sectores productivos. COLCIENCIAS, Bogotá, pp. 14

Coleman, J. (1990): Foundations of social theory. Belknap Press

Cooke, P. and Morgan, K. (1994): Growth regions and duress. Renewal strategies in Baden-Württemberg and Emilia-Romagna. In: Ash, A. and Trift, N. (1994): Holding down the global: Possibilities for local economic policy. Oxford University Press, New York, pp. 91-117

Cooke, P. (1996): The wave of regional innovation networks. Analysis, characteristics and strategy. Small Business Economics, Vol. 8, pp. 159-171

Cooke, P. (1998): Introduction. Origins of the concept. In: Braczyk, H. et al. (eds.) (1998): Regional innovations systems. The role of governance in a globalized world. UCL Press Ltd., pp. 2-25

Cooper, C. (1980): Policy interventions for technological innovation in developing countries. World Bank Staff Working Paper No. 441, pp. 59

CORPEI (2002a): Perfil de producto: Bambú (*Guadua angustifolia*). Proyecto CORPEI-CBI Expansión de la oferta exportable de Ecuador. CORPEI, pp. 31

CORPEI (2002b) : Reporte Bambú No. 1/2002

CORPEI (2002c): Reporte Bambú No. 2/2002

CORPOCALDAS (2002) : Microcluster de la Guadua. Corporación Atonoma Regional de Caldas and the Camara de Comercio de manizales. Brochure, pp. 42

CORTES, C. AND MONTOYA, J. (1986) : Aspectos económicos en la producción y comercialización de la Guadua (*Guadua angustifolia* Kunth) en el departamento de Caldas. Tesis de grado. Facultad de Agronomía. Universidad de Caldas, Manizales, pp. 82

CRESPO & ASSOCIADOS (2001) : Estudio de mercado. Muebles de bambú. Mercado Estados Unidos. Crespo & Associados. Marketing Agency. Guayaquil, Ecuador, pp. 64

CROUCH, C. (2003): Institutions within which real actors innovate. In: MAYNTZ, R. AND STREECK, W. (eds.) (2003): Die Reformierbarkeit der Demokratie. Innovationen und Blockaden. Festschrift für Fritz W. Scharpf. Campus Verlag, pp. 71-98

DAGILIS, T. AND WIT, H. (2003): Bamboo value-added export development. Oportunities for Ecuador. Report prepared for CORPEI, Ecuador, pp. 82

DEBEER, J. AND MCDERMOTT, M. (1989): The economic value of non-timber forest products in South-East Asia with emphasis on Indonesia, Malaysia and Thailand. IUCN, Amsterdam

DEEKE, A. (1995): Experteninterviews – ein methodologisches und forschungspraktisches Problem. In BRINKMANN, C., DEEKE, A. AND VÖLKEL, B. (1995): Experteninterviews in der Arbeitsmarktforschung. Diskussionsbeiträge zu methodischen Fragen und praktischen Erfahrungen. Institut für Arbeitsmarkt- und Berufsforschung Nürnberg, pp. 7-22

DERAS, E. (2003): Análisis de la cadena productiva del bambú en Costa Rica. Tesis de Maestria en Socioeconomia Ambiental. CATIE, Costa Rica, pp. 126

DILGER, G. (1996): Kolumbien. Beck`sche Länder Reihe, pp. 146

DOLOM, P. (2003): Criteria and indicators for assessing the sustainability of a community based forest management project in the Philippines. Unasylva, Vol. 54, 214/215, pp. 22-29

DOSI, G., PAVITT, K. AND SOETE, L. (1990): The economics of technical change and international trade. Harvester Wheatsheaf, pp. 303

DRESCHER, R. (2002): Von der Uruguay-Runde zur Ministererklärung von Doha. Was gibt es Neues? E+Z 36, No. 6, pp. 9-13

DUMONT, M. AND TSAKANIKAS, A. (2001): Knwoledge spillovers through R&D networking. Paper prepared for the OECD, pp. 29

DURST, P., ULRICH, W. AND KASHIO, M. (eds.) (1994): Non-wood forest products in Asia, FAO RAPR Publication 1994/28, Bangkog

EBELING, W. AND SCHARNHORST, A. (2002): Technische Evolution aus der Sicht der Theorie der Selbstorganisation. PARTHEY, H. AND SPUR, G. (eds.) (2002): Wissenschaft und Innovation. Wissenschaftsforschung Jahrbuch 2001. GEWIF, Berlin, pp. 224

ELIASSON, G. (2000): Industrial policy, competence blocs and the role of science in economic development. Journal of Evolutionary Economics, Vol. 10, pp. 214-241

ELIASSON, G. (2001): The role of knowledge in economic growth. In: HRDC/OECD. International Symposium Report September 2001, pp. 42-64

ENDRES, A. AND QUERNER, I. (1993): Die Ökonomie natürlicher Ressourcen. Eine Einführung. Wissenschaftliche Buchgesellschaft, pp. 173

ESEGU, F., SSENTEZA, J. AND SEKATUBA, J. (2000): Rattan and bamboo in Uganda. A study of the production-to-consumption systems. INBAR Working Paper Series No. 29, pp. 39

ESPINAL, C. AND ROLDAN, D. (2000): La concertatción publico – privada: Los acuerdos de competitividad en Colombia. Consorcio técnico del IICA. Area de politicas y commercio. IICA, pp. 78

ESSER, H. (2002): Was könnte man (heute) unter einer Theorie mittlerer Reichweite verstehen? In: MAYNTZ, R. (ed.) (2002): Akteure, Mechanismen, Modelle. Zur Theoriefähigkeit makro-sozialer Analysen. Campus Verlag, pp. 128-150

EßER, K. ET AL. (1992): Neue Determinanten internationaler Wettbewerbsfähigkeit. Erfahrungen aus Lateinamerika und Ostasien. DIE

EßER ET AL. (1995a): Systemische Wettbewerbsfähigkeit. Neue Anforderungen an Unternehmen und Politik. Vierteljahresheft zur Wirtschaftsforschung. Jhrg. 64. Nr. 2, pp. 186-199

EßER, K. ET AL. (1995b): Neue Tendenzen der Exportförderung. Anforderung an Konzept und Instrumente. Beispiel Uruguay, DIE

EßER, K. ET AL. (1999): Systemische Wettbewerbsfähigkeit und Entwicklung. In: THIEL, R. (ed.): Neue Ansätze zur Entwicklungstheorie. DSE Themendienst 10. ZD, Bonn, pp. 147-163

FAO (1998): Asia-Pacific forestry towards 2010. report of the Asia-Pacific forestry sector outlook study. FAO, pp. 242

FAO (1999): State of the world's forests 1999. FAO, pp. 154

FAO (2001): State of the world's forests 2001. FAO, pp. 181

FAO (2001b): Global forest resources assessment 2000. Main report. FAO, pp. 479

FAUSTMANN, M. (1849): Berechnung des Werthes, welchen Waldboden, sowie noch nicht haubare Holzbestände für die Waldwirtschaft besitzen. Allgemeine Forst- und Jagdzeitung, Dezember 1849, pp. 441-455

FEDER, G. (1981): Adoption of agricultural innovations in developing countries. A survey. World Bank Staff Working Paper No. 444, pp. 67

FERRANTI, D. ET AL. (2002): From natural resources to the knowledge economy. The World Bank Latin American and Caribbean studies, pp. 178

FERRANTI, D. ET AL. (2003): Closing the gap in education and technology. The World Bank Latin American and Caribbean Studies, pp. 203+Bibliography

FERROUKHI, L. (ed.) (2003): Municipal forest management in Latin America. CIFOR/IDRC

FINGER, A (1999): Metals from the forest. Mining and forest degradation. IUCN/WWF, pp. 40

FLICK, U. (1996): Qualitative Forschung. Theorie, Methoden, Anwendung in Psychologie Und Sozialwissenschaften. Rowohlts Enzyklopädie, pp.318

FORERO, J. (1999): Economia y sociedad rural en los Andes Colombianos. COLCIENCIAS, Colombia, pp. 378

FORERO, J. ET AL. (2002): Sistemas de producción rurales en la región Andina Colombiana. Análisis de su viabilidad económica, ambiental y cultural. Colciencias, Colombia, pp. 234

FREEMAN, C. (1974): The economics of industrial innovation. Penguin Books, pp. 409

FREEMAN, C. (1987): Technology policy and economic performance. Lessons from Japan. Frances Pinter, pp. 155

FUHR, H. (1993): Induzierter institutioneller und organisatorischer Wandel. Theoretische Fundamente und Transformationserfahrungen am Beispiel lateinamerikanischer Entwicklungsländer. Universität Konstanz, Fakultät für Verwaltungswissenschaft, pp. 348

FUHR, H. (1998): Staatsreform and Verwaltungsmodernisierung. Zur neuen Rolle des Staates in Lateinamerika. Universität Potsdam, Wirtschafts- und sozialwissenschaftliche Fakultät. Papier zur Publikation, pp. 25

GABLER, S. (1992): Schneeballverfahren und verwandte Stichproben. ZUMA-Nachrichten No.31, pp. 47-69

GABLER WIRTSCHAFTSLEXIKON (1993): L-O. Gabler Verlag, pp. 2021-2527

GIRALDO, E. AND SABOGAL, A. (eds.) (1999): Una alternative sostenible: La Guadua. Corporación Autonoma Regional del Quindio (CRQ), Colombia, pp. 192

GONZALES, E. AND POSADA, A. (1996): Estudio general de la situación actual y potencial de la Guadua (*Guadua angustifolia* Kunth). Subregion Urabá Antioqueñio. Universidad Nacional de Colombia, sede Medellin, pp. 107

GORDO, M. (2002): El cumplimiento tributario voluntario en Colombia. Notas para la discusión. United Nations Online Network in Public Administration and Finance (UNPAN), pp. 16

GRANOVETTER, M. (1973): The strength of weak ties. American Journal of Sociology, Vol. 78, pp. 1360-1380

GRANOVETTER, M. (1978): Threshold models of collective behaviour. American Journal of Sociology, Vol. 83, No. 6, pp. 1420-1443

GRANOVETTER, M. (1982): The strength of weak ties. A network theory revisited. In: MARDSEN, P.AND LIN, N. (eds.): Social structure and network analysis. Sage Publishers, pp. 105-129

GRANOVETTER, M. (1992): Economic action and social structure. The problem of embeddedness. In: GRANOVETTER, M. AND SWEDBERG, R. (1992): The sociology of economic life, Westview, pp. 53-81

GRANOVETTER, M. AND SOONG, R. (1983): Threshold models of diffusion and collective behaviour. Journal of Mathematical Sociology, Vol. 9, pp. 165-179

GRANOVETTER, M. AND SOONG, R. (1986): Threshold models of interpersonal effects in consumer demand. Journal of economic Behaviour and Organization, Vol. 7, pp. 83-89

GRANOVETTER, M. AND SOONG, R. (1989): Threshold models of diversity. Chinese restaurants, residential seggragation and the spiral of silence. Sociological Methodology, Vol. 18, pp. 69-104

GTZ (1997): Waldschutz durch Kleingewerbe. GTZ, Eschborn, pp. 35

GTZ (2001a): Exemplarische Darstellung von Ansätzen der lokalen Wirtschaftsförderung in Lateinamerika. GTZ, Eschborn, pp. 80

GTZ (2001b): Drogen und Entwicklung in Lateinamerika. Strategien, Erfahrungen und Projektbeispiele aus der Arbeit der GTZ. Aktionsprogramm Drogen und Entwicklung/GTZ, pp. 70+Annex

GUAIPATIN, C. (2002): Agroindustrial clusters in Latin America. What explains their resistance? Paper presented at the international workshop on "Fostering rural economic development through agriculture-based enterprises and services", held from 20.11.-22.11.2002 in Berlin. GTZ, pp. 10

GUNDLACH, E. (1999): Humankapital als Motor der Entwicklung. Ein neuer Ansatz der neoklassischen Wachstumstheorie. In: THIEL, R. (ed.): Neue Ansätze zur Entwicklungstheorie. DSE Themendienst 10. ZD, Bonn, pp. 173-185

HAGEN, J. (1997): Agrarian interests in a liberalizing economy. Colombia's law 101 of 1993. Paper presented at the meeting of the Latin American Studies Association, held in Guadalajara, Mexico, April 1997, pp. 21

HAHN, H. AND PREUß, H. (1994): Regionalplanung in der ländlichen Entwicklung. Materialien des Zentrums für regional Entwicklungsforschung, Universität Gießen, Band 31, pp. 118

HANUMAPPA, H. (1996): Agarbathi. A bamboo based industry in India. INBAR Working Paper Series No. 9, pp. 22

HARTMANN, G. AND HÄRTER, F. (1989): Allgemeine Wirtschaftslehre für kaufmännische Auszubildende. Merkur Verlag Rinteln, pp. 472

HARBORTH, H. (1993): Sustainable development. Dauerhafte Entwicklung. In: NOHLEN, D. AND NUSCHELER, F. (eds.) (1993): Handbuch der dritten Welt 1. Grundprobleme, Theorien, Strategien. Verlag J.H.W. Dietz Nachf., pp.231-247

HAUSER, E. (2002): WTO, Ernährungssicherung und Armutsbekämpfung, ein Widerspruch? E+Z 36, No. 6, pp. 14-17

HEADLEY, M. (2003): Participatory forest management. The Jamaica Forestry Department experience.
Unasylva, Vol. 54, 214/215, pp. 44-50

HELD, C. AND MANZANO, I. (2004): La cadena productiva y el mercado regional de la Guadua en el Eje Cafetero Colombiano. INBAR Working Paper No. 51, pp. 61

HEMMER, H. AND WILHELM, R. (1999): Neue Hoffnung für Entwicklungsländer? Entwicklungspolitische Implikationen endogener Wachstumstheorien. In: THIEL, R. (ed.): Neue Ansätze zur Entwicklungstheorie. DSE Themendienst 10. ZD, Bonn, pp. 249-270

HERRERA, D. (1999): Metodologia para la elaboración de tipologias de actores. Serie cadenas y dialogo para la acción. IICA Area de politíca y commercio, pp. 97

HERRERA, D. (2000): Competitividad con equidad en cadenas agroalimetarias. Consorcio técnico del IICA. Area de politicas y commercio. IICA, pp. 52

HIDALGO, O. (2003): Bamboo. The gift of the gods. Oscar Hidalgo, pp. 553

HILLEBRAND, W. (1991): Industrielle und technologische Anschlussstrategien in teilindustrialisierten Ländern. Bewertung der allokationstheoretischen Kontroverse und Schlussfolgerungen aus der Fallstudie Republik Korea, DIE

HOOD, C. (1998): The art of the state. Culture, rhetoric and public management. Oxford University Press

HOUSE ET AL. (2002): Understanding culture and implicit leadership theories across the globe. An introduction to project GLOBE. Journal of World Business Vol. 37, pp. 3-10

HUMBOLDT, VON A. (1821): Personal narratives of travels to the equinoctial regions of the new continent during the years of 1799-1804. Longman UK

HUMPHREY, J. AND SCHMITZ, H. (2000): Governance and upgrading. Linking industrial cluster and global value chain research. IDS Working Paper 120, pp. 37

HYTTINEN, P., MONONEN, A. AND PELLI, P. (eds.) (1996): Regional development based on forest resources. Theories and practices. EFI Proceedings No. 9, pp. 265

INBAR (1999): INBAR´s assessment of socioeconomic issues and constraints in the bamboo and rattan sectors. INBAR, pp. 28

JANSSEN, J. (1998): The wood of the poor. The relationship between bamboo, the environment and the local community. Gate Vol. 1/98. pp. 26-30

JIAFU, L. (2001): A development strategy for bamboo resources and industry in China. In: ZHAOHUA, Z. (ed.) (2001): Sustainable development of the bamboo and rattan sectors in tropical China. INBAR Proceedings No. 6. China Forestry Publishing House, pp. 1-18

JUDZIEWICZ, E. ET AL. (1999): American bamboos. Smithsonian Institution Press, Washington and London, pp. 392

KÄRKKÄINEN, M. (1996): How to develop forest based industry. In: HYTTINEN, P., MONONEN, A. AND PELLI, P. (eds.) (1996): Regional development based on forest resources. Theories and practices. EFI Proceedings No. 9, pp. 65-70

KALMANOVITZ, S. (2001a): Las instituciones y el desarrollo económico en Colombia. Grupo Editorial Norma, pp. 302

KALMANOVITZ, S. (2001b): Las instituciones Colombianas en el siglo XX. Alfaomega,

Libros de Cambio, pp. 51

KANT, S AND CHIU, M. (2000): Bamboo sector reforms and the local economy of Linan county, Zhejiang province, People's Republic of China. Forest, Policy and Economics, Vol. 1/2000, pp. 283-299

KAPLINSKY, R AND READMAN, J. (2001): Integrating SMEs in global value chains. Towards partnership for development. UNIDO, Vienna, pp. 82

KAPPEL, R. (1999): Endogene Potentiale und die Ökonomie der Peripherie. Handels- und raumtheoretische Ansätze. In: THIEL, R. (ed.): Neue Ansätze zur Entwicklungstheorie. DSE Themendienst 10. ZD, Bonn, pp. 271-284

KARKI, M., SHERCHAN, G. AND KARKI, J. (1998): Extensive production-to-consumption system in Eastern Nepal. A case study. INBAR Working Paper Series No. 17, pp. 41

KELBESSA, E. ET AL. (2000): A socio-economic case study of the bamboo sector in Ethiopia. An analysis of the production-to-consumption system. INBAR Working Paper Series No. 25, pp. 41

KELLE, U. AND ERZBERGER, C. (2000): Qualitative und quantitative Methoden: Kein Gegensatz. In: FLICK, U. (Ed.) (2000): Qualitative Forschung. Ein Handbuch. Rowohlts Enzyklopädie, pp. 299-309

KELLE, U. (2001): Sociological explanations between micro and macro and the integration of qualitative and quantitative methods. Forum Qualitative Sozialforschung Vol. 2, No. 1, 2001 (Online Journal). 43 paragraphs

KERN, H. AND SCHUMANN, M. (1984): Das Ende der Arbeitsteilung? Rationalisierung in der industriellen Produktion. Beck, pp. 361

KLEINING, G. AND WITT, H. (2001): Discovery as basic methodology of qualitative and quantitative research. Forum Qualitative Sozialforschung Vol. 2, No. 1, 2001 (Online Journal). 81 paragraphs

KNORRINGA, P. (1999): Agra. An old cluster facing the new competition. World Development, Vol. 27, No. 9, pp. 1587-1604

KÖSSLER, R. (1998): Entwicklung. Reihe Einstiege. Verlag Westfälisches Dampfboot, pp. 227

KÖSTERS, W. (1994): Neue Wachstumstheorie und neue Außenhandelstheorie. Frische Argumente für eine staatliche Industriepolitik. Wirtschaftswissenschaftliches Studium H 3, pp. 117-122

KOECHLIN, G. (2000): Characterizing cluster innovation. An approach from the wood products sector. Department of Economics and Finance, Universidad de Talca, Chile, pp. 8

KOGLER, A. (1991): Investitionen in Produkt- und Prozessinnovationen. Eine evolutionsökonomische Analyse der Struktur und Entwicklung des Innovationsmix zwischen Produkt- und Prozessinnovationen. Peter Lang Verlag, pp. 219

KOSCHATZKY, K. (2001): Räumliche Aspekte im Innovationsprozess. Ein Beitrag zur neuen Wirtschaftsgeografie aus Sicht der regionalen Innovationsforschung. LIT-Verlag, pp. 437

KOWOL, U. AND KROHN, W. (2000): Innovation und Vernetzung. Die Konzeption der Innovationsnetzwerke. In: WEYER, J. (ed.) (2000): Soziale Netzwerke. Konzepteder sozialwissenschaftlichen Netzwerkforschung. Oldenbourg Verlag, pp. 135-160

KOYO, J. (2003): Final statement of the XII World Forestry Congress. Envisaging a new future for the forests. Unasylva, Vol. 54, No. 214/215, p. 13

KRAAY, A., SOLOAGA, I. AND TYBOUT, J. (2002): Product quality, productive efficiency, and international technology diffusion. THE WORLD BANK. Policy Research Working paper No. 2759, pp. 46

KRINGS, T. AND MÜLLER, B. (2001): Politische Ökologie. Theoretische Leitlinien und aktuelle Forschungsfelder. In: REUBER, P. AND WOLKERSDORFER G. (eds.) (2001): Politische Geographie. Handlungsorientierte Ansätze und Critical Geopolitics. Heidelberger Geographische Arbeiten. Heft 112, pp. 93-116

KRUGMAN, P. (1991): Geography and trade. Leuven University Press, pp. 142

LAI, Q. (2003): Community participation in the management of nature reserves. Experiences and lessons form China. Unasylva, Vol. 54, 214/215, pp. 51-58

LALL, S. (1992): Technological capabilities and industrialization. World Development, Vol. 20, No. 2, pp. 165- 186

LACHMANN, W. (2003): Volkswirtschaftslehre 1. Grundlagen. Springer-Verlag, pp. 313

LEDERMAN, D. AND MALONEY W. (2003): R&D and development. Background paper prepared for FERRANTI ET EL. (2003). The World Bank Latin American and Caribbean Office, pp. 37

LENTZ, C. (1992): Quantitative und qualitative Erhebungsverfahren im fremdkulturellen Kontext. Kritische Anmerkungen aus ethnologischer Sicht in: REICHERT, C. ET AL (1992): Empirische Sozialforschung über Entwicklungsländer. Methodenprobleme und Praxisbezug. Kölner Beiträge zur Entwicklungsländerforschung. Verlag Breitenbach

LEVITSKY, J. (1993): Innovations in the financing of small and microenterprises in developing countries. Small enterprises development working paper. SED 22/E. International Labour Organization (ILO), pp. 93

LEVIN, D. AND CROSS, R. (2002): The strength of weak ties you can trust. The mediating role of trust in effective knowledge transfer. Paper proceeding of the Academy of Management, pp. 34

LIESE, W. (1998): The anatomy of bamboo culms. INBAR, Technical Report, pp. 205

LOAIZA, I., MONTENEGRO, W. AND OLAYA, J. (2001): Industrialización del bambú. Tesis de Maestria en Administración de Empresas. Escuela superior politécnica del Litoral, Guayaquil, Ecuador, pp. 56

LOASBY, B. (2000): Market institutions and economic evolution. Journal of Evolutionary Economics, Vol. 10, pp. 297-309

LONDOÑO, X. (1992): La Guadua – Un bambú importante de America. Revista Procana. No.

46, pp. 10-14

LONDOÑO, X. (2001): Evaluation of bamboo resources in Latin America. INBAR Working Paper Series No. 32, pp. 30

LÓPEZ, C. (1999): Formulación plan de administración de los rodales de Guadua existentes en el municipio de Pereira, departamento de Risaralda. Tesis de grado. Facultad de Ciencias ambientales. Universidad Tecnologica de Pereira, pp. 111

LÓPEZ, M. AND SIERRA, D. (2000): Identificación de oportunidades empresariales en Risaralda a partir del recurso Guadua. Tesis de grado. Facultad de Ingenieria industrial. Universidad Tecnologica de Pereira, pp. 120

LÓPEZ, A. (2001): Guadua. Alternativas de comercialización internacional. CARDER, Pereira, pp. 78+Annex

LUCAS, R. (1988): On the mechanics of economic development. Journal of Monetary Economics, 22, pp. 3-42

LÜCKE, M. (1992): The diffusion of process innovations in industrialized and developing countries. A case study from the world textile and steel industries. Kiel Working paper No. 535. Institute for World Economics, Kiel, pp. 22

LUHMANN, N. (1994): Soziale Systeme. Grundriss einer allgemeinen Theorie. Suhrkamp Wissenschaft, pp. 675

LUJAN, C. ET AL. (2003): Sustainable forest development in Mexico. A hierarchical system of criteria and indicators. Unasylva, Vol. 54, 214/215, pp. 30-36

LUNDVALL, B. (1992): National systems of innovation. Towards a theory of innovation and interactive learning. Pinter Publishers, pp. 342

MACHADO, A. (1998): La cuestión agraria en Colombia a fines del milenio. El Ancora Editores, Bogotá, pp. 222

MACHADO, A. (2002): De la estructura agraria al sistema agroindustrial. Universidad Nacional de Colombia. Sede Bogotá, pp. 320

MÄKINEN, P. AND SELBY, A. (1996): The success factors of forest and woodworking SMEs. In: HYTTINEN, P., MONONEN, A. AND PELLI, P. (eds.) (1996): Regional development based on forest resources. Theories and practices. EFI Proceedings No. 9, pp. 71-84

MAGGI, C. (2000): Key factors of structural change in North-Rhine Westphalia, INEF Report No. 45, Duisburg

MAOGONG, Z. ET AL. (1998): Bamboo in Anji. A case study of an intensive production-to-consumption system. INBAR Working Paper Series No. 15, pp. xx

MANSKE, F. (in prep.): Theorie der Innovation

MANSKE, F. AND MOON, Y. (1998): Innovation auf Koreanisch. Die Entwicklung EDI-Systemen in der koreanischen und deutschen Automobilindustrie. Research Report

MANSKE, F., RUTH, K. AND DEITMER, L. (2000): Sustaining innovation networks by empowerment? Recent findings of innovation and evaluation research as starting points for the development and application of an actor-centred evaluation method. Institut für Technik und Bildung, Universität Bremen, pp. 19

MATHEW, P. (1998): The bamboo economy of Kerala, India. An analysis of the production-to-consumption systems. INBAR Working Paper Series No. 12, pp. xx

MAYNTZ, R. (1991): Modernization and the logic of inter-organizational networks. Max-Planck-Institut für Gesellschaftsforschung, Köln

MAYNTZ, R. (1993): Große technische Teilsysteme und ihre gesellschaftstheoretische Bedeutung . Kölner Zeitschrift für Soziologie und Sozialpsychologie, 45, pp. 97-108

MAYNTZ, R. (ed.) (2002): Akteure, Mechanismen, Modelle. Zur Theoriefähigkeit makro-sozialer Analysen. Campus Verlag, pp. 235

MAYNTZ, R AND SCHARPF, F. (1995): Der Ansatz des akteurszentrierten Institutionalismus. In: MAYNTZ, R. AND SCHARPF, F. (eds.) (1995): Gesellschaftliche Selbstregelung und politische Steuerung, Campus Verlag, pp. 368

MAYRING, P. (1993): Einführung in die qualitative Sozialforschung. Beltz Psychologie Verlags Union, pp. 132

MAYRING, P. (2000): Qualitative Inhaltsanalyse. Forum Qualitative Sozialforschung Vol. 1, No. 2, 2000 (Online Journal). 28 paragraphs

MAYRING, P. (2001): Kombination und Integration qualitativer und quantitativer Analyse. Forum Qualitative Sozialforschung Vol 2, No. 1, 2001 (Online Journal). 31 paragraphs

MCKELVEY, M. (1998): Evolutionary innovations. Learning, entrepreneurship and the dynamics of firms. Journal of Evolutionary Economics, Vol. 8, pp. 157-175

MEJIA, F. (1952): La Guadua. Tésis de Ingenieria agricola. Universidad Nacional de Colombia, sede Medellin, pp. 29

MELLOR, J. AND DESAI, M. (1985): Agricultural change and rural poverty. IFPRI/Johns Hopkins University Press, pp. 233

MENZEL, U. (1993): 40 Jahre Entwicklungsstrategie = 40 Jahre Wachstumsstrategie. In: NOHLEN, D. AND NUSCHELER, F. (eds.) (1993): Handbuch der dritten Welt 1. Grundprobleme, Theorien, Strategien. Verlag J.H.W. Dietz Nachf., pp. 131-155

MERTEN, K. (1995): Inhaltsanalyse. Einführung in Theorie, Methode und Praxis. Westdeutscher Verlag, pp. 406

MESSNER, D. (1990): Von der Importsubstitution zur weltmarktorientierten Spezialisierung. Optionen für den Industriesektor Uruguays. DIE

MEYER-STAMER, J. (1991): „Perestroquinhia". Neuorientierung in der brasilianischen Industriepolitik. Lateinamerika – Analysen, Daten, Dokumentation, Nr. 16, 1991

MEYER-STAMER, J. (1996a): Technologie und industrielle Wettbewerbsfähigkeit. Allgemeine Überlegungen und Erfahrungen aus Brasilien. Weltforum Verlag, pp. 301

MEYER-STAMER, J. (1996b): Industriepolitik für den Industriestandort Deutschland. Jenseits des japanischen Erfolgsmodells. Friedrich-Ebert-Stiftung, Bonn, pp. 200

MEYER-STAMER, J. (1999a): Strategien lokaler/regionaler Entwicklung. Cluster, Standortpolitik und systemische Wettbewerbsfähigkeit. Projekt Meso-NRW. Institut für Entwicklung und Frieden. Universität Duisburg, pp. 25

MEYER-STAMER, J. (1999b): Lokale und regional Standortpolitik. Konzepte und Instrument jenseits von Industriepolitik und traditioneller Wirtschaftsförderung. INEF Report 39/1999. Institut für Entwicklung und Frieden, Universität Duisburg, pp. 31

MEYER-STAMER, J. (2000) : Meso-Laboratorium Nordrhein-Westfalen. Beobachtungen zur Struktur- und Standortpolitik in einer altindustriellen Region. INEF Report No. 47, Duisburg

MEYER-STAMER, J. (2001): Was ist Meso? Systemische Wettbewerbsfähigkeit: Analyseraster, Benchmarking-Tool und Handlungsrahmen. Projekt Meso NRW Report 55/2001, pp. 48

MEYER-STAMER, J. (2001b) : Systemische Wettbewerbsfähigkeit. Vom zufälligen Konzept zum Benchmarking-tool. In: ALTENBURG, T. AND MESSNER, D. (2001): Wettbewerbsfähiges Lateinamerika. Herausforderungen für Wirtschaft, Gesellschaft und Staat, DIE

MEYER-STAMER, J. (2002): Clustering and the creation of an innovation-oriented environment for industrial competitiveness. Beware of over optimistic expectations. Paper prepared for the international high level seminar on technological innovation, held from 05.09.-07.09.2000 in Beijing, China, pp. 23

MEYER-STAMER, J. AND SEIBEL, S. (2002): Cluster, value chain and the rise and decline of collective action. The case of the tile industry in Santa Catarina, Brazil. Institute for Development and Peace, University of Duisburg, pp. 22

MEYER-STAMER, J. ET AL. (1996) : Industrielle Netzwerke und Wettbewerbsfähigkeit. Das Beispiel Santa Catarina/Brasilien. DIE, Berlin

MIEG, H. (1994a): Die Expertenrolle. UNS Working Paper. ETH Zürich. Human Environmental Interactions (MUB), pp. 18

MIEG, H. (1994b): Experteninterviews. MUB Working paper No. 6. ETH Zürich. Human Environmental Interactions (MUB), pp. 21+11

MORAN, J. (2001): Usos tradicionales y actuales del bambú en América Latina, con énfasis en Colombia y Ecuador. Available from INBAR and the Centro de investigaciones territoriales del Ecuador, pp. 192

MORRILL, R. (1968): Waves of spatial diffusion. Journal of Regional Science, Vol. 8, pp. 1-18

MOROSINI, P. (2003): Industrial clusters, knowledge integration and performance. World Development, Vol. 32, No. 2, pp. 305-326

MOSER, C. (2000): Violence in Colombia. Building sutainable peace and social capital. In: SOLIMANO, A. (ed.) (2000): Colombia. Essays on conflict, peace and development. The World Bank, pp. 9-77

MSBC (2001): Norma uificada en Guadua. Reglamentación para el manejo, aprovechamiento y establecimiento de Guadua, Caña brava y bambú. Proyecto MSBC (Manejo Sostenible de Bosques en Colombia), Bogotá, pp. 16

MSBC (2003) : Términos de referencia unificados para estudios y planes de manejo y aprovechamiento forestal de la Guadua. Proyecto MSBC (Manejo Sostenible de Bosques en Colombia), Bogotá, pp. 40

NADVI, K. (1999): Collective efficiency and collective failure. The response of the Sialkot surgical instrument cluster to global quality pressure. World Development, Vol. 27, No. 9, pp. 1605-1626

NAVARRO, G. (2003): On 189 years of confusing debates over the KÖNIG-FAUSTMANN formula. Schriften aus dem Institut für Forstökonomie der Universität Freiburg, Band 18, pp. 221

NELSON, R. (ed.) (1993): National innovation systems. A comparative analysis. Oxford University Press

NELSON, R. (2002): Bringing institutions into evolutionary growth theory. Journal of Evolutionary Economics, Vol. 12, pp. 17-28

NELSON, R. AND WINTER, S. (1975): Growth theory from en evolutionary perspective. The differential productivity growth puzzle. American Economic review, Vol. 65, pp. 338-344

NELSON, R. AND WINTER, S. (1977): In search of a useful theory of innovation. Reseach Policy, Vol. 6, pp. 36-76

NELSON, R. AND WINTER, S. (1982): An evolutionary theory of economic change. Harvard University Press, pp. 437

NIELSEN, K. (2000): Social capital and systemic competitiveness. Research Papers in Network Institutional Theory No. 2/00, pp. 46

NIELSEN, K. (2003): Social capital and innovation policy. Research Papers in Network Institutional Theory No. 10/03, pp. 29

NOHLEN, D AND NUSCHELER, F. (1993a): Was heißt Entwicklung? In: NOHLEN, D. AND NUSCHELER, F. (eds.) (1993): Handbuch der dritten Welt 1. Grundprobleme, Theorien, Strategien. Verlag J.H.W. Dietz Nachf., pp. 55-75

NOHLEN, D. AND NUSCHELER, F. (1993b): Indikatoren von Unterentwicklung und Entwicklung. In: NOHLEN, D. AND NUSCHELER, F. (eds.) (1993): Handbuch der ^ dritten Welt 1. Grundprobleme, Theorien, Strategien. Verlag J.H.W. Dietz Nachf., pp. 76-108

NON-WOOD NEWS (2003): Bamboo in China. Vol. 10/2003, p. 8

NON-WOOD NEWS (2004): Products and markets. Bamboo. Vol. 11/2004, pp. 25-28

NORTH, D. (1988): Theorie des institutionellen Wandels. Eine neue Sicht der Wirtschaftsgeschichte. Mohr, Tübingen, pp. 228

NORTH, D. (1991): The new institutional economics and third world development. In: Harris, J., Hunter, J. and Lewis, C. (eds.) (1995): The New Institutional economics and third world development. Routledge, New York, pp. 17-26

NORTH, D. (1993): The New Institutional Economics and development. Economic Working Paper Archive. Economic History Series, No. 9309002, pp. 8

NUSCHELER, F. (1991): Lern- und Arbeitsbuch Entwicklungspolitik. Verlag J.H.W. Dietz Nachf., pp.348

NUSCHELER, F. (1993): Entwicklungspolitische Bilanz der 80er Jahre. Perspektiven für die 90er Jahre. In: NOHLEN, D. AND NUSCHELER, F. (eds.) (1993): Handbuch der dritten Welt 1. Grundprobleme, Theorien, Strategien. Verlag J.H.W. Dietz Nachf., pp. 156-178

OECD (1996): Oslo Manual. The measurement of scientific and technological activities. Proposed guidelines for collecting and interpreting technological innovation data. EUROSTAT/European Commission, pp. 92

OGLIASTRIE, E. (1998): Culture and organizational leadership in Colombia. Paper prepared at the Universidad de los Andes, Bogota, pp. 43

OHLIN, B. (1921): Concerning the question of the rotation period. Reprinted in: Journal of Forest economics (1995) Vol. 1, No. 1, pp. 89-114

OLIVER, P. AND MARWELL, G. (1988): The paradox of group size in collective action. A theory of the critical mass II. American Sociological Review, Vol. 53, pp. 1-8

OLSON, M. (1998): Die Logik des kollektiven Handelns. Kollektivgüter und die Theorie der Gruppe. Mohr Siebeck Verlag, pp. 181

ONGUGO, P. ET AL. (2000): Production-to-consumption-systems. A case study from the bamboo sector in Kenya. INBAR Working Paper Series No. 27, pp. 53

OSPINA, S. (2000) : Cómo mejorar el estatus de la Guadua en la economía del país. Revista Ciencias Humanas de la Universidad Tecnologica de Pereira. No. 25, pp. 120-124

PERÉZ, M., SAYER, J. AND COHEN, S. (1993): El extrativismo en America Latina. Conclusiones y recomendaciones del taller UICN-CCE, Amacayacu, Colombia, Ocubre 1992

PERÉZ, M. AND BYRON, N. (1998): A methodology to analyze divergent case studies of non-timber forest products and their development potential. Forest Science, Vol. 45, No. 1/1999, pp. 14

PÉREZ, M. ET AL. (1999): The role of bamboo plantations in rural development. The case of Anji county, Zheijang, China.World Development, Vol. 22 No. 1/99, pp. 101-114

PÉREZ, M. ET AL. (2001): Forestry, poverty and rural development in China. Some views from the bamboo sub-sector. Paper presented at the international symposium: Lessons learned from the Chinese forest policy experience, held in Dujiangyan, China, June, 2001, pp. 24

PÉREZ, O. (1980): Reforestación con Guadua. Una posibilidad economica. Revista ESSO Agricola. Vol. 27, No. 1, pp. 5-8

PETERS, C. ET AL. (1989): Valuation of an Amazonian rainforest. Nature, No. 339, pp. 655-656

PICKENPACK, L. (2004): Innovation in der Forstwirtschaft. Freiburger Schriften zur Forst- und Umweltpolitik, Band 5, pp. 239

PIZANO, D. (2001): El café en la encrucijada. Evolución y perspectivas. Alfaomega, Libros de Cambio, pp. 80

PLATTEAU, J. (1994a): Behind the market stage. Where real societies exist. Part I: The role of public and private order institutions. Journal of Development Studies, Vol. 30, No. 3, pp. 533-577

PLATTEAU, J. (1994b): Behind the market stage. Where real societies exist. Part II: The role of moral norms. Journal of Development Studies, Vol. 30, No. 4, pp. 753-817

POPPER, K.D. (1976): Logik der Forschung. Tübingen, pp. 441

PORTER, M. (1990): The competitive advantage of nations. New York Free Press, pp. 855

POSADA, A. (1991): Perspectivas de producción y comercialización de la Guadua en el oriente y suroeste Antioqueño. Tesis de grado. Facultad de Ciencias humanas. Universidad Nacional de Colombia, sede Medellin, pp. 264+Annex

POTI, B. AND BASILE, R. (2000): Regional differences of innovation. Firms` organizational, regional institutions and innovative performance. Paper prepared for the OECD-NIS meeting: Focus Group on Innovative Firms and networks held in Rome, Italy from 2nd to 3rd of October 2000, pp. 44

PRAHL, R., MARWELL, G. AND OLIVER, P. (1991): Reach and selectivity as strategies of recruitment for collective action. A theory of critical mass V. Joutnal of Mathematical Sociology, Vol. 16, No. 2, pp. 137-164

PÜTTMANN, W. (2002): Einführung. In: Forstliche Entwicklungshilfe (2002). Schriftenreihe der Landesforstverwaltung Baden-Württemberg. Band 80, pp. 428

PYKE, F., BECATTINI, G. AND SENGENBERGER W. (1990): Industrial district and firm co-operation in Italy. International Labour Organisation, Geneva

RABELLOTTI, R. (1999): Recovery of a Mexican cluster. Devaluation bonanza or collective efficiency. World Development, Vol. 27, No. 9, pp. 1571-1586

REPUBLICA DE COLOMBIA (1999). MINISTERIO DEL MEDIO AMBIENTE: Politicas ambientales de Colombia. Imprenta Nacional, pp. 610

REPUBLICA DE COLOMBIA (2001). MINISTERIO DE COMERCIO EXTERIOR. Exportación de pisos en Guadua a la U.E. MINCOMEX/PROEXPORT, pp. 60

REPUBLICA DE COLOMBIA (2002). CONTRALORÍA. Modelo, política e institucionalidad agropecuaria y rural. Imprenta Nacional, Colombia, pp. 174

REPUBLICA DE COLOMBIA (2002). MINISTERIO DE AGRICULTURA: VILLABA, R. (Autor): Informe al Congreso 1998-2002. Reactivación y modernización con visión de futuro. Imprenta Nacional, pp. 338

REPUBLICA DE COLOMBIA. DIAN (2004). (DIRECCIÓN DE IMPUESTOS Y ADUANAS NACIONALES DE COLOMBIA). www.dian.gov.co

REPUBLICA DE COLOMBIA. LEY NO. 590 DE JULIO DE 2000: Por la cual se dictan disposiciones para promover el desarrollo de las micro, pequeñas y medianas empresas

REPUBLICA DE COLOMBIA (2003). MINISTERIO DE PROTECCIÓN SOCIAL: Resolución No. 0474 DE 2003 of the 3rd of April 2003, pp. 2

REPUBLICA DE COLOMBIA (2004). DEPARTMENTO ADMINISTRATIVO NACIONAL DE ESTADISTICA (DANE): www.dane.gov.co. Internet database enquiry, March 2004

REVISTA NACIONAL DE AGRICULTURA (2001a): Importancia del sector agropecuario en la economia Colombiana. No. 931 Tercer trimestre, 2001, pp. 8-25

REVISTA NACIONAL DE AGRICULTURA (2001b): El sector agropecuario en cifras. No. 931 Tercer trimestre, 2001, pp.69-72

RICARDO, D. (1817): On the principles of political economy and taxation. Online book at the Library of Economics and Liberty (www.econlib.org)

RINCÓN, O. (1977): La Guadua y su importancia socioeconómica. Revista ESSO Agricola, Vol. 24, No. 3, pp. 5-12

RÍOS, H. (1994): La Guadua: Nuestro bambú. CRQ, Armenia, pp. 137

ROELANDT, T., GILSING, V. AND VAN SINDEREN, J. (2000): New policies for the new economy. Cluster based innovation policy. International experiences. Paper presented at the 4th annual EUNIP conference held in Tilburg, Netherlands, from 07th to 09th of December 2000, pp. 25

ROGERS, E. (1995): Diffusion of innovations. Free Press New York

ROMER, D. (2001): Advanced macro-economics. McGraw-Hill Higher Education, pp. 651

ROMER, P. (1986): Increasing returns and long-run growth. Journal of Political Economy, 94, pp. 1002-1037

ROMER, P. (1990): Endogenous technological change. Journal of Political Economy, 98, pp. 71-102

RUÍZ, M. AND AMELINES, W. (2002): Prefactibilidad técnica financiera para una planta de transformación primaria de Guadua en el municipio de Cocorná. Tesis de grado. Facultad nacional de Minas. Universidad Nacional de Colombia, sede Medellin, pp. 200+Annex

RYAN, R AND GROSS, N. (1943): The diffusion of hybrid seed corn in two Iowa communities. Rural Sociology, Vol. 8, No. 1, pp. 15-24

SAMPER-KUTSCHBACH, M. (1995): In difficult times: Colombian and Costa Rican coffee growers from prosperity to crisis, 1920-1936. In ROSEBERRY, W. ET AL. (eds.) (1995): Coffee, society and power in Latin America, pp. 151-180

SCHARPF, F. (1993): Positive und Negative Koordination in Verhandlungssystemen. In: HERITIER, A. (1993): Policy-Analyse. Kritik und Neuorientierung. Politische Vierteljahresschrift, Sonderheft Nr. 24, Verlag Opladen, pp. 57-94

SCHARPF, F. (2000): Interaktionsformen. Akteurzentrierter Institutionalismus in der Politikforschung. UTB, pp. 474

SCHIFF, M. AND WANG, Y. (2002): Regional integration and technological diffusion. The case of the NAFTA. Paper prepared for International Trade Unit of THE WORLD BANK, pp. 20

SCHIMANK, U. (2002): Handel und Strukturen. Einführung in die akteurtheoretische Soziologie. Grundlagentexte Soziologie. Juventa Verlag, pp. 366

SCHMIDT, U. (2002): Der Wald in Deutschland im 18. und 19. Jahrhundert. CONTE Forst Verlag, pp. 434

SCHMIDT, V. (2003): The boundaries of „Bounded Generalizations". Discourse as the missing factor in actor-centred institutionalism. In: MAYNTZ, R. AND STREECK, W. ^ (eds.) (2003): Die Reformierbarkeit der Demokratie. Innovationen und Blockaden. Festschrift für Fritz W. Scharpf. Campus Verlag, pp. 318-350

SCHMITHÜSEN, F. ET AL. (2003): Unternehmerisches Handeln in der Wald- und Forstwirtschaft. Betriebswirtschaftliche Grundlagen und Managementprozesse. dbv, pp. 560

SCHMITZ, H. (1995): Collective efficiency. Growth path for small-scale industry. Journal of Development Studies, Vol. 31, No. 4, pp. 529-566

SCHMITZ, H. (1999): Global competition and local co-operation. Success and failure in the Sinos Valley, Brazil. World Development, Vol. 27, No. 9, pp. 1627-1650

SCHNEIDER, B. (1998): The state and collective action. Business politics in Latin America. Paper presented at the meeting of the Latin American Studies Association in Chicago, USA, September 1998, pp. 40

SCHNEIDER, V. AND MAYNTZ, R. (1995): Akteurzentrierter Institutionalismus in der Technikforschung. Fragestellungen und Erklärungsansätze. In: HALFMAN, J. ET AL. (eds.) (1995): Technik und Gesellschaft Jahrbuch 8, pp. 107-130

SCHNELL, R., HILL, P. AND ESSER, E. (1995): Methoden der empirischen Sozialforschung. Oldenbourg Verlag, pp. 506

SCHUMPETER, J. (1934): Theorie der wirtschaftlichen Entwicklung. Duncker und Humblot, Berlin (reprinted in 1987), pp. 369

SCOBIE, G. AND JACOBSEN, V. (1995): Structural adjustment and agriculture. In: TABOR, S. (ed.) (1995): Agricultural research in an era of adjustment. The World Bank, pp. 3-21

SCOTT, A. (2000): Economic geography. The great half century. Cambridge Journal of Economics, Vol. 24, pp. 483-504

SEIBEL, H. (Eds.) (1992): Empirische Sozialforschung über Entwicklungsländer. Methodenprobleme und Praxisbezug. Verlag Breitenbach Publishers, pp. 317-339

SILVERMAN, D. (2000): Interpreting qualitative data. Methods for analysing talk, text and interaction. Sage Publications, pp. 325

SINGH, M., BERSALONA, C. AND QUINTANS, K. (2000): Bamboo in Abra (Philippines). An investigation of the production-to-consumption system. INBAR Working Paper Series No. 30, pp. 48

SMITH, A. (1776): An inquiry into the nature and causes of the wealth of nations. Online book at the Library of Economics and Liberty (www.econlib.org)

SOLIMANO, A. (1999): Globalization and national development at the end of the 20th century. Tensions and challenges. The World Bank Policy Research Working Paper No. 2137, pp. 14

SOLOW, R. (1956): A contribution to the theory of economic growth. Quarterly Journal of Economics 70 (February), pp. 65-94

SOLOW, R. (1970): Growth Theories. An exposition. Clarendon Press Oxford, pp. 107

SRIVASTAVA, B. (1998): Management and utilization of bamboo and rattan in Papua New Guinea. INBAR Working Paper Series No. 18, pp. 32

STAMM, J. (2002): Informa final: Proyecto latas y laminados de Guadua. Proyecto UTP-GTZ. Facultad de Ciencias ambientales. Universidad Tecnologica de Pereira, pp. 33

STATZ, J. (2000): Nutzung von Nicht-Holz-Waldprodukten als Handlungsfeld der Entwicklungszusammenarbeit. Freiburger Forstliche Forschung. Band 11, pp. 235

STATZ, J. (2000b): Entwicklungspotentiale der Nutzung von Nicht-Holz-Waldprodukten. Perspektiven für ein neues Feld der forstlichen Entwicklungszusammenarbeit in Paraguay und Bolivien. GTZ/TÖB, Eschborn, pp. 106

STOIAN, D. (2000): Shifts in forest product extraction. The post-rubber era in the Bolivian Amazon. Dissertation and der Universität Freiburg, Institut für Forst- und Umweltpolitik, Arbeitsbereich Markt & Marketing

STORPER, M. (1997): The regional world. Territorial development in a global economy. The Guilford Press, pp. 338

SUAREZ, R. (2001): Los desafios rurales en Colombia. Elementos institucionales del exito. Colciencias/TM Editores, pp. 214

SUDMAN, S. AND KALTON, G. (1986): New Developments in the Sampling of Special Populations. Annual Review of Sociology 12, pp. 401-429

TABOR, S. (1995): Structural adjustment and institutional change. In: TABOR, S. (ed.) (1995): Agricultural research in an era of adjustment. The World Bank, pp. 23-35

TAFORI (TANZANIAN FORESTRY RESEARCH INSTITUTE) (2000): Bamboo production-to-consumption systems in Tanzania. INBAR Working Paper Series No. 28, pp. 45

TAINTER, J. (1988): The collapse of complex societies. Cambridge University Press, pp. 250

TEMU, A. (1995): Assessment of non-timber resources In Africa. In: KÖHL, M. ET AL. (eds.) (1995): The Monte Verita conference on forest survey designs. Simplicity versus efficiency and assessment of non-timber resources, held from 02.05.-04.05.1994 in Ascona, pp. 110-115

THE WORLD BANK (2001): World Development Report 2000/2001. Attacking poverty. Oxford University Press, pp. 335

THE WORLD BANK (2002): Colombia poverty report. Vol. I, pp. 112

THE WORLD BANK (2003): Breaking the conflict trap. Civil war and development policy. World Bank Policy Research Report, pp. 211

THIEL, R. (1999): Zur Neubewertung der Entwicklungstheorie. In: THIEL, R. (ed.): Neue Ansätze zur Entwicklungstheorie. DSE Themendienst 10. ZD, Bonn, pp. 9-34

THOROE, C. AND OLLMANN, H. (1998): Entwicklung internationaler Wettbewerbsfähigkeit der deutschen Forst- und Holzwirtschaft. In: SEKOT, W. (ed.) (1998): Beiträge zur Forstökonomik. Band 31. Festschrift für o. Univ. Prof. Dr. Wolfgang Sagl. Institut für Sozioökonomik der Forst- und Holzwirtschaft, Wien, pp. 243-264

UNMÜSSIG, B. AND WALTHER, M. (1999): Strukturanpassung in Afrika. Warten auf den Post-Washington Konsensus. E+Z 40, No. 9, pp. 247-250

URREA, F. ET AL. (2000): Innovación y cultura de las organizaciones en tres regiones de Colombia. Colciencias, pp. 308

VANTOMME, P. (2003): Can forests be sustainably managed for non-wood forest products? Unasylva, Vol. 54, No. 214/215, pp. 43

VERBECK, H. (1999): Innovative clusters. Identification of value-adding production chains and their networks of innovation. An international study. Doctoral thesis, Erasmus University, Amsterdam, pp. 141

VILLARONGA-WALKER, B. (1998): Das Konzept der nachhaltigen Entwicklung und seine Umsetzung in Kolumbien. Magisterarbeit der Universität Mainz, Fachbereich Sozialwissenschaften, pp. 174+11

WARDLE, P. (2002): Report on the establishment of a database of statistics on the trade in bamboo and rattan. ITTO, pp. 48

WEBER, M. (1922): Wirtschaft und Gesellschaft. In: WEBER, M. (1979): Gesammelte Aufsätze zur Wirtschaftslehre. Mohr Verlag Tübingen, pp. 579

WEI, F. AND CHI-SON, C. (1997): China's bamboo industry. Tropical Forest Update Vol.7, No. 4/97, p.15

WEIZSÄCKER VON, R. (1990): Technik und Natur. In: SCHÜZ, M. (1990): Risiko und Wagnis. Neske Verlag, pp. 14-25

WEYER, J. (ed.) (2000): Soziale Netzwerke. Konzepte der sozialwissenschaftlichen Netzwerkforschung. Oldenbourg Verlag, pp. 317

WHITEMAN, A., BROWN, C. AND BULL, G. (1999): Forest product market developments. The outlook for forest product markets to 2010 and the implications for improving management of the global forest estate. working paper prepared for the THE WORLD BANK, FAO, available online: www.fao.org

WILLIAMS, M. (2001): The history of deforestation. Online Journal History Today, July 2001 (www.historytoday.com)

WILLIAMS. M. (2003): Deforesting the earth. From prehistory to global crisis. University of Chicago Press, pp. 689

WILLIAMSON, O. (1995): Institutions and economic organization. The governance perspective. The World Bank, Washington

WILLKE, H. (1988): Staatliche Intervention als Kontextsteuerung Kritische Vierteljahresschrift für Kriminologie und Gesetzgebung, 1988, pp. 214-228

WOMACK, J., JONES, D. AND ROOS, D. (1990): The machine that changed the world. Rawson, New York

WUNDER, S. (2003): Oil ealth and the fade of the forest. A comparative study of eight countries. Routledge Explorations in Environmental Economics, pp. 432

ZHANG, Y. ET AL. (1999): The forest sector in China. Towards a market economy. In: PALO, M. AND UUSIVUORI, J. (eds.) (1999): World forests, society and environment. Kluwer Academic Publishers Dordrecht, pp. 371-393

ZHAOHUA, Z. (ed.) (2001): Sustainable development of the bamboo and rattan sectors in tropical China. INBAR Proceedings No. 6. China Forestry Publishing House, pp. 260

Annex 1: Maps

Map 1: Colombia with study region

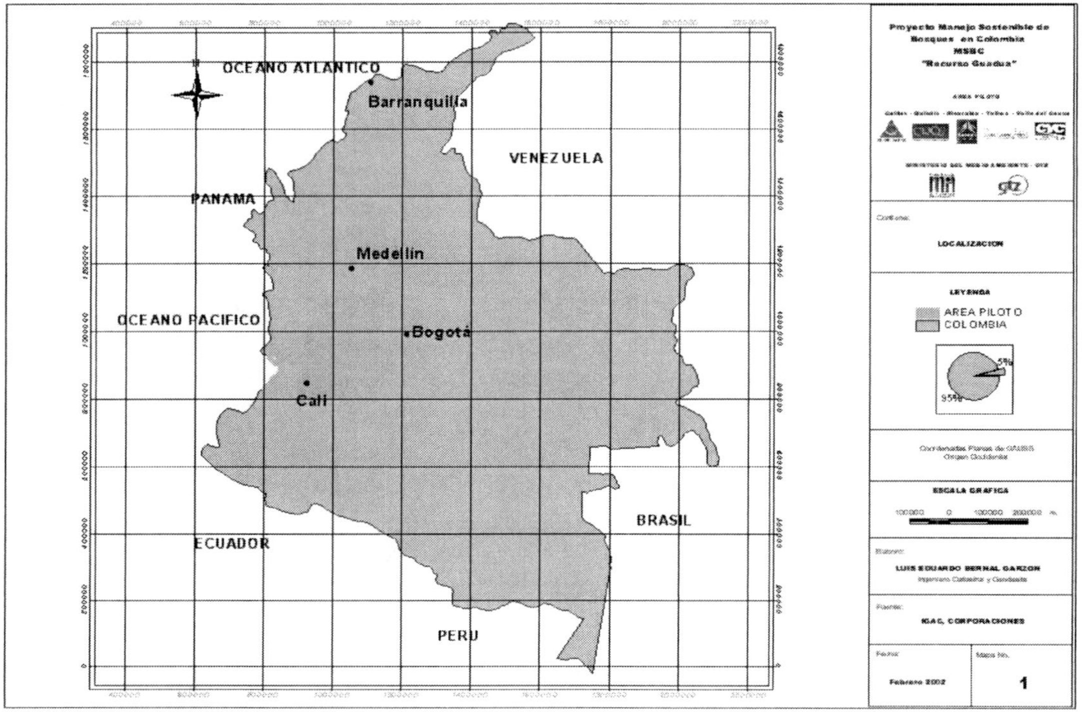

Map 2: Study Region

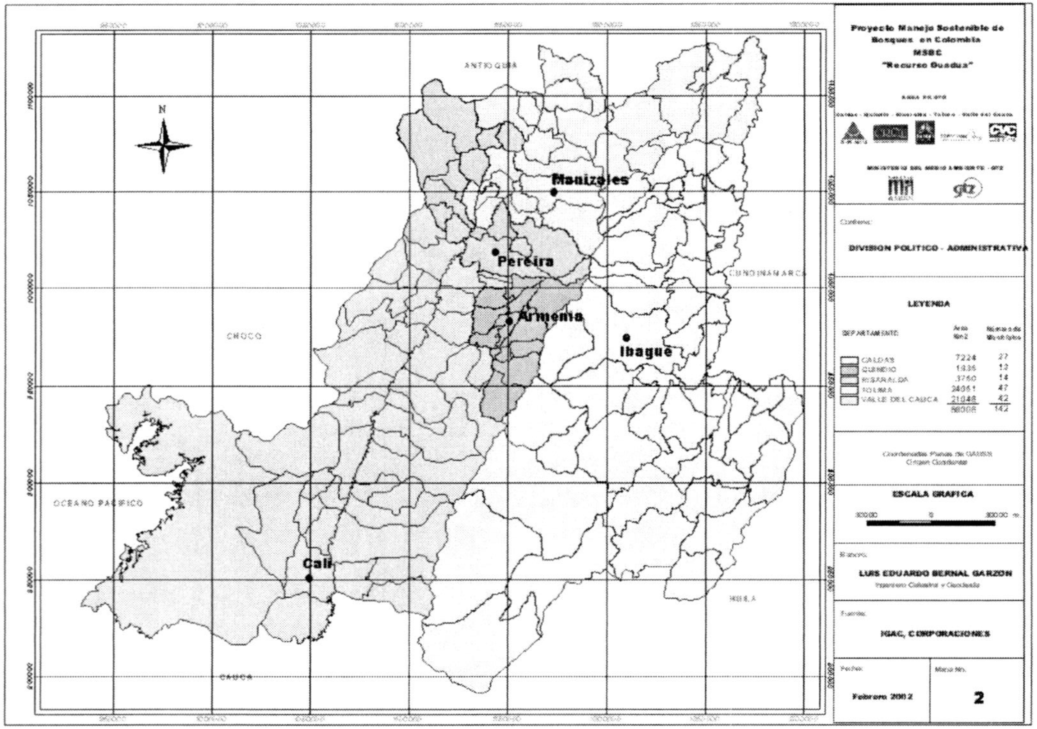

Annex 2: Questionnaires

Questionnaire 1: Meso-level Actors

Cuestionario: Inventario y analisis de actores e instituciones

Nombre del encuestado : Funcción/posición del encuestado : Organización/Proyecto/Empresa : Dirección :

Sección I : Desarrollo regional

Qué es, en su opinión, desarrollo regional ?
Cuáles son los fines del desarrollo regional para el eje cafetero ?
Cuáles son las mayores contribuciones de su organización con respecto al desarrollo regional y la Guadua ?

Sección II : Recursos y capacidades de los actores

Describa las fuerzas y las debilidades de su organización con respecto al desarrollo regional.
Qué mas va su organización hacer para mejorar su participación en este proceso ?

Sección III : Estrategias e interacciones de los actores

Cómo influye su organización en la toma de decisiones en el proceso de desarrollo regional ?
Dónde ve usted la clave para un desarrollo satisfactorio con respecto a la situación en general y con respecto a la Guadua ?
Con quién interactúa su organización en este proceso ?
Hay otras vias de comuncación que son de importancia ?

Sección IV : La Guadua

En su opinión, qué son los productos o aplicaciones más prometedores para el desarrollo regional (Por qué ?) ?
Cómo promover estos productos ?
Cómo contribuye su organización en el desarrollo o la promoción de estos productos ?
Hay un grupo de interese especial, que son los productores pequenos de la Guadua. Comó se promueve este grupo ?

Sección V : Preguntas del encuestado

Tabla 1: Auto-Evaluación organizacional

Inversiones de recursos en ...	Ano (2007 = Proyección)	Rango (1-5)	Ejemplos	Resultados	
				Ano	Evalua-ción (0-2)
Capacitación economica	1997			1997	
	2002			2002	
	2007			xxxxxxxxxxxxxxx	
Capacitación técnica	1997			1997	
	2002			2002	
	2007			xxxxxxxxxxxxxxx	
Investicación sistemas de producción	1997			1997	
	2002			2002	
	2007			xxxxxxxxxxxxxxxxx	
Formación técnica	1997			1997	
	2002			2002	
	2007			xxxxxxxxxxxxxxx	
Investigación biologica	1997			1997	
	2002			2002	
	2007			xxxxxxxxxxxxxxx	
Investigación mercado	1997			1997	
	2002			2002	
	2007			xxxxxxxxxxxxxxx	
Desarrollo de productos	1997			1997	
	2002			2002	
	2007			xxxxxxxxxxxxxxx	
Desarrollo de lineas de producción	1997			1997	
	2002			2002	
	2007			xxxxxxxxxxxxxxx	
Participación en asociaciones o gremios	1997			1997	
	2002			2002	
	2007			xxxxxxxxxxxxxxx	
Formación institucional	1997			1997	
	2002			2002	
	2007			xxxxxxxxxxxxxxx	
Subvención de proyectos/ Confinanziació	1997			1997	
	2002			2002	
	2007			xxxxxxxxxxxxxxx	
Publicidad/ Ferias	1997			1997	
	2002			2002	
	2007			xxxxxxxxxxxxxxx	

Instrucciones Tabla 1:

1) Escoge sectores en donde su organización/proyecto/empresa esta activa y evalúe los rangos ségun su importancia de 1 a 5 (1 = peor importancia ; 5 = mayor importancia). Definición de recursos : Recursos de su organización/proyecto/empresa son recursos financieros, técnicos, personales, institucionales.
2) Despues evalúe los resultados de las inversiones de su recursos en rangos de 0 a 2 (0 = no satisfactorio ; 1 = satisfactorio ; 2 = mas que satisfactorio).
3) Haga ejemplos para los sectores (qué tipo de recursos para qué ?) y resultados escogidos (razon para su evaluación).

Tabla 2: Interacciones

Con cuáles otros actores interactúa su organización/proyecto/empresa con respecto al desarrollo regional y al desarrollo del sector Guadua ? Cuáles son los objetivos de estos interacciones y existen resultados ? Cómo ve usted el futuro de las interacciones con estos actores ?

Actor	Ano de iniciación	Objetivos de las interaciones	Resultados de las interacciones	Proyección

Tabla 3: Evaluación de otros actores

Evalue otros actores con respecto a sus inversiones de recursos en el sector Guadua y los resultados conseguidos con estas inversiones.
0 = no satisfactorio ; 1 = satisfactorio ; 2 = mas que satisfactorio

Inversion de recursos/Actor															
Financiero	0	1	2	0	1	2	0	1	2	0	1	2	0	1	2
Personal	0	1	2	0	1	2	0	1	2	0	1	2	0	1	2
Institucional	0	1	2	0	1	2	0	1	2	0	1	2	0	1	2
Tecnologico	0	1	2	0	1	2	0	1	2	0	1	2	0	1	2
Comentarios															

Ejemplos de sectores para tabla 1 :

Capacitación empresarial	Talleres
	Seminarios
	Insumos
Capacitación técnica	Talleres
	Seminarios
	Insumos, Sitios
Formación profesional	Formación profesional con/sin examen/graduación
	Practicás/Pasantias
	Formación profesional empresarial/organizacional
Investigación biologica	Laboratorios
	Proyectos
	Estudios
Investigación mercado (Vista cliente)	Proyectos
	Estudios/Sondeos
	Sistema de información
Desarrollo de productos (Vista product.)	Investigación técnica en productos
	Laboratorios
	Estudios/Sondeos
Desarrollo de lineas de producción	Investigación técnica (Maquinária)
	Talleres/Laboratorios/Fabricas
	Transferencia de tecnologia
Participación en asociaciones o gremios	Participación en gremios o asociaciones regularmente
	Participación en gremios o asociaciones no regularmente
	Funcción en estos gremios/asociaciones
Formación institucional	Contribución en toma de deciciones
	Formación de normas o leyes ; o de vias de comuniación
	Asesoria en estos processos
Subvención de proyectos	Asistencia financaria a otros organizaciones/proyectos
	Asistencia técnica a otros organizaciones/proyectos
	Asistencia institucional a otros organizaciones/proyectos
Publicidad	Publicidad/Promoción
	Publicación de resultados/Conferencias/Ferias

Questionnaire 2: Handicraft and Furniture Producer Questionnaire 2002

Cuestionario: Análisis del sistema de producción a consumo (PCS)
Sector Artesanal y Muebles

Nombre del encuestado:
Función/posición del encuestado:
Tipo de empresa:
Dirección de la empresa:

1. Cuándo comenzó usted con su negocio y cuándo empezó a comercializar/procesar productos asociados con la guadua? Si el negocio es reciente, qué los motivó a iniciar el negocio?

2. Adquisición de productos de guadua/materia prima

Tipo de producto	Precio por unidad	Cantidad/t	Provededor	Lugar (Dónde?)
Rizomas				
Cepas (5m)				
Cepas (6m)				
Sobrebasas				
Varillon				
Ramas				
Otros				

2.a) Cómo paga usted las compras ?

3. Productos más comprados de Guadua

Producto	Cantidad de producción/t	Precio de venta	Cantidad de venta/t

3 a. Otros materiales empleados/combinados con la guadua (costos de producción)

Producto	Guadua (%)	Otro 1 (%)	Otro 2 (%)	Otro 3 (%)

4. Tratamiento de la Guadua

Parte de la Guadua	Producto	Proceso	Tiempo	Empleados	Precio
Rizomas					
Cepas (5m)					
Cepas (6m)					
Sobrebasas					
Varillon					
Ramas					
Otros					

5. Tecnología:
Describa las diferentes tecnologías que usted aplica en su empresa (negocio) (1) en general y (2) para el caso específico de la guadua.

Maquina	Guadua u otra aplicación	Quién la produce	Dónde	Precio

6. Venta y utilización de maderas más importantes

6.a Generalidades

Nombre	Tipo de Productos	Precio	Aplicaciones	De dónde proviene?

6.b Dinámica de las ventas, precios y tipos de maderas y aplicación en los últimos años.

Comentarios:

7. Calidad del Producto

a) Qué criterios de calidad aplica usted cuando compra productos de guadua?

Colores	Diámetros	Longitudes	Parte de la guadua	Tratamiento posterior	Otros (Cuáles?)

b) Cómo chequea usted la calidad de dichos productos?

Sentidos	Maquinaria	Tecnología	Otros (Cuáles?)

c) Hay entregas de mala calidad y para que se utiliza?

Si	No	Algunas veces

d) Paga usted diferentes precios según la variación de la calidad?

Si	No	Algunas veces

e) Cuáles son las deficiencias más comunes en la calidad de los productos de guadua/guadua?

Rajaduras	Humedad	Malformaciones	Defectos visuales	Otros (Cuáles?)

f) Qué criterios de calidad aplican sus compradores?

Colores	Dimensiones	Durabilidad	Parte de la guadua	Tratamiento posterior	Otros (Cuáles?)

g) Qué tipo de calidad es el más comprado/deseado en su negocio?

Optima	Buena	Media	Baja	Dominancia del precio

8. Situación del mercado:

a) Existen insuficiencias en la cantidad de material para algún pedidoy tiene que rechazar pedidos debido a la insuficiencia en la cantidad de material, deficiencias de calidad o falta de capacidad de su empresa?

Si	No	Algunas veces

b) Son sus provedores puntuales y confiables en cuanto a la calidad y cantidad del material?

Si	No	Algunas veces

c) Qué tipo de compradores tiene usted?

Permanentes	Regulares	Ocasionales	Observaciones

d) Tiene usted pedidos en grandes cantidades?

Permanentemente	Ocasional	Regularmente	Cantidades				Otros (Cuáles)
			10+	50+	100+	500+	

e) Aplicación de estas cantidades. Cuál es la finalidad de estas cantidades compradas?

Cantidad y producto				

9. Estructura de la empresa:

a) Tiene usted empleados o socios? Cuantos ?

(1-5)	(6-10)	(11-15)	(16-20)	Más de 20

b) Cómo les paga?

Por unidad de producto	Por día trabajado	Mensualmente	Según volúmen de ventas	Otros (Cuáles?)

c) Qué tipos de cualidades tienen?

Ebanistas (carpinteros)	No preparados	Otros (Cuáles?)

d) Ha variado el número de empleados/socios en los últimos diez años?

1992	2002	Observaciones

10. Volúmen de Ventas:

a) Cuál es el volúmen total anual de ventas?

Total (mio. $)					Parte de Guadua (%)	Otra 1	Otra 2
5-10	10-20	20-30	30-50	50 o más			

b) Cómo se ha comportado el volumen de ventas de su empresa y en especial el volúmen de ventas de la guadua en los últimos años?

Comentario:

c) Cómo se han comportado los precios de los productos de guadua en los últimos años?

Tipo de producto	2002	1992

11. Instituciones:

a) Que regulaciones o leyes del gobierno conoce usted; conoce su influencia sobre el uso y mercadeo de la guadua? Y Quién impone dichas normas?

Ley/Regulación	Entidad responsable	A quién importa ?	Observaciones

b) Interactúa usted con esas organizaciones/entidades ? Por qué?

Si	No	Algunas veces	Observaciones

c) Paga cargas o retribuciones por la comercialización/procesamiento de la guadua? Paga usted impuestos ?

Tipo	Si/No	Algunas veces	Observaciones
Cargas			
Retribuciones			
Impuestos			

d) Hay otras regulaciones oficiales/informales que afecten el sector guadua?

Descripción	Observaciones

e) Esas regulaciones promueven u obstaculizan el sector guadua?

Promueven	Obstaculizan	Observaciones

12.Gremios:

a) Hay gremios o asociaciones que representen sus intereses? Usted participa ?

Si	Cuáles?	Part.	No	Observaciones

b) Quiénes se benefician con esas asociaciones?

Comentarios:

c) Tiene usted contacto con cualquier otra organización relacionada con la guadua?

Si	Cuáles?	No	Observaciones

d) Asiste o ha asistado usted a convenios o talleres sobre la Guadua

e) Asiste a ferias o exposiciones ? Porqué ?

Feria/Expos.	Dónde ?	Cuándo ?	Productos	Resultados

13. Cooperación y competición:

a) Coopera usted con otros comerciantes, productores, precesadores u otros ?

Quién ?	Si/No	Objetivos/Por qué no ?
Productores		
Comerciantes		
Procesadores		
Otros		

b) Quién ofrece los mismos productos que usted ?

c) Había antes mas competición ?

14. Disponibilidad de los créditos bancarios, subsidios u otros promociones economicos ?

a) Tiene usted acceso a subsidios o créditos bancarios? para qué?

Si	No	Algunas veces	Observaciones (para qué?)

b) Dónde?

Comentario:

c) Cuáles son las condiciones de los créditos/subsidios ?

Comentario:

d) Está usted afectado por cualquier crédito relacionado con el sector guadua?

Comentario:

e) Si usted no tiene acceso a créditos o subsidios, cuál es la causa de dicha situación?

Comentario:

15. El Futuro:

a) Planea para el futuro algún cambio en la estructura de su empresa?

Si	No	Tal vez	Observaciones

b) Planea algún cambio referente al negocio de la guadua?

Si	No	Tal vez	Observaciones

16. Preguntas y comentarios del encuestado

Questionnaire 3:
Handicraft and Furniture Producers 2004 and Questionnaire for innovative Producers of Lamps 2004

I. CUESTIONARIO INNOVADORES DE ARTESANÍAS

1. Qué tipo de empresa posee? Continúa con su empresa?

Si su respuesta es **SI**:

- Su negocio ha presentado crecimiento?

Año	# de Empleados	Volumen de ventas
2003		

- Producen nuevos productos?
- Qué productos?
- Vale la pena producirlos? Por qué los producen?
- Dónde aprendió a producir ese producto?
- Ha comprado nueva tecnología?

Tipo de Tecnología	Producto Fabricado	Forma de Pago		
		Financiación	Capital Propio	Otro

- Ha habido cambios en la estructura organizacional de su empresa?

Nuevos Socios	Nuevo tipo de Empresa	Otros

Si su respuesta es **NO**:

- Por qué?
- A qué se dedica ahora?

II. PRODUCTORES DE LÁMPARAS

Nombres:

Dirección:

Identificación:

1. Desde qué año produce lámparas?

2. Dónde aprendió a hacer lámparas? El contacto con esa persona/organización fue regular u ocasionalmente?

Jefe Anterior	Otro Artesano	Capacitación	Autodidacta	Otro

3. Usted ha aprendido a producir un tipo especial de lámpara?

4. Qué tipo de modificaciones ha hecho para mejorar el producto o hacer la producción más fácil?

Diseño	Accesorios	Colores	Materiales	Otros

5. Tiene empleados? Hubo empleados que actualmente tienen su propio negocio y producen lámparas?

# de empleados	Con negocio propio	Producen Lámparas

Nombres:

Dirección:

6. Qué piensa de esa situación?

Calidad	Diseño	Precio	Otros

7. Hay otros artesanos que copian sus lámparas?

Nombres:

Dirección:

8. Hace un año, usted dijo que las lámparas son un producto muy importante para su negocio:

- Como están las cosas ahora?
- Vale la pena producir lámparas todavía?
- Qué ha cambiado?
- Hay más competencia?

11. Tiene productos inventados por usted?

Producto	Es exitoso?	Por qué?	Observaciones

12. Quisiera producir productos de otros artesanos?

13. Por qué no está produciendo esos productos?

14. Respecto a los productos nuevos o lámparas:

- Cuáles fueron los costos más importantes al comienzo?
- Usted ha invertido tiempo y dinero en esos productos?
- Hubo problemas

Conocimientos Técnicos	Legislación	Otros

Annex 3: Detailed list of interviewed meso-level actors in 2002

Actor Group	Number	Details
Government Entities	13	• Ministry of Foreign Trade • Ministry of Agriculture • Ministry of Environment • 4 Regional Environmental Authorities (CARs) • 2 Regional Centres of Extension and Education (SENA) • The National Centre for Bamboo Research • 3 Regional Centres for the Promotion of Bio-commerce
NGOs	4	• 2 regional PACOFOR Offices (funded by the Netherlands) • 1 regional PROCUENCA Office (funded by the FAO) • CIPAV (Colombian NGO)
Construction Sector Assoc.	5	• National Association of seismic Engineering • National Construction Chamber, regional Office • National Organisation for Standardisation • National Architecture Society, regional Office • National Society of civil Engineering, regional Office
Agricultural Producer Orgs.	3	• National Association of the Sugarcane Producers and Processors • National Association of the Coffee Growers • National Coffee Research Centre
Research Entities	7	• 4 public universities • 3 private universities
Public-private Actor Associations	5	• The National Laboratory for Design • Regional Development Foundation of the Quindio • The Bamboo Convention • 2 Regional Business Chambers
Independent Experts	6	Anonymous

Annex 4: Table of answer items of guideline interviews according to the qualitative contents analysis

Analytical level	Meso-level			Meso- and micro-level			Meta-level	
Answer Category	Intra-organisational problems	Inter-organisational problems	Problems in decision-making processes	Institutional barriers to sector development	Deficits in involvement and interaction	General sector problems	Socio-cultural problems	Security problem
Cognitional Items In brackets: % of interviewees stated this item during the guideline interview (n=43)	Lack of financial resources (30%) Lack of thematic continuity (19%) Lack of skilled/knowledgeable personal (14%) Personality bounded thematic agenda (14%) Lack of technical equipment (7%) Problems in intra-organisational communication (7%)	Problems in inter-organisational communication (54%) Lack of willingness to co-operate (51%) Missing publication of research and work results (35%) Lack of co-ordination of activities (35%) Lack of research co-ordination (23%) Occupation of spaces without adequate qualification (7%)	Lack of consideration of research and work results (28%) Lack of participation/ Top-down decision-making (19%) Lack of interaction with municipalities (16%) Complex and long-term decision-making process/lack of transparency (14%)	Institutional definition problems ("NTFP-Problem") (33%) Lack of loan possibilities/barriers to access credits (28%) Lack of incentives for producers and SMEs (19%) Lack of resource control/management (19%) Lack of export promotion (16%) High transaction costs (14%) Protectionist resource management (12%)	Lack of private activities/investments in the sector (56%) Lack of public activities/investments in the sector (40%) Lack of co-operation/integration with/of private actors (26%) Lack of co-operation/integration with/of research entities (19%) Occupation of spaces/topics by inadequate actors (7%)	Institutional obstacles (65%) Lack of organisation of target groups (42%) Investment resource distribution problems (35%) Lack of sector organisation (35%) Lack of public resources (26%) Infrastructure obstacles (12%)	Lack of valuation of the resource and its products (75%) Undefined cultural obstacles (56%) Lack of co-operational traditions/ Mistrust (35%) Lack of entrepreneurial culture (23%) Social heterogeneity (19%)	The security situation in Colombia was perceived as an obstacle to investments in the economy, transport and trade and as negatively affecting the attractiveness of Colombia for foreign investors. Further, it was seen as reason for the low degree of trust based co-operation. (30%)

Annex 5: Cross table of Meso-level actor groups and the institutional background groups of their co-operations partners

Anzahl

		Partner Group							
		overnmenta	NGO	onstruction	griculture	esearch	ublic-private Actor	Others	Total
Actor Group	Governmental	24	0	0	1	7	10	8	50
	NGO	8	1	0	0	3	0	0	12
	Construction	6	0	1	1	0	1	2	11
	Agriculture	6	0	0	0	0	0	1	7
	Research	9	1	0	0	2	2	2	16
	Public-private A	13	1	0	2	3	10	1	30
Total		66	3	1	4	15	23	14	126

Annex 6: Self-evaluation of the meso-level actors 1997 and 2002

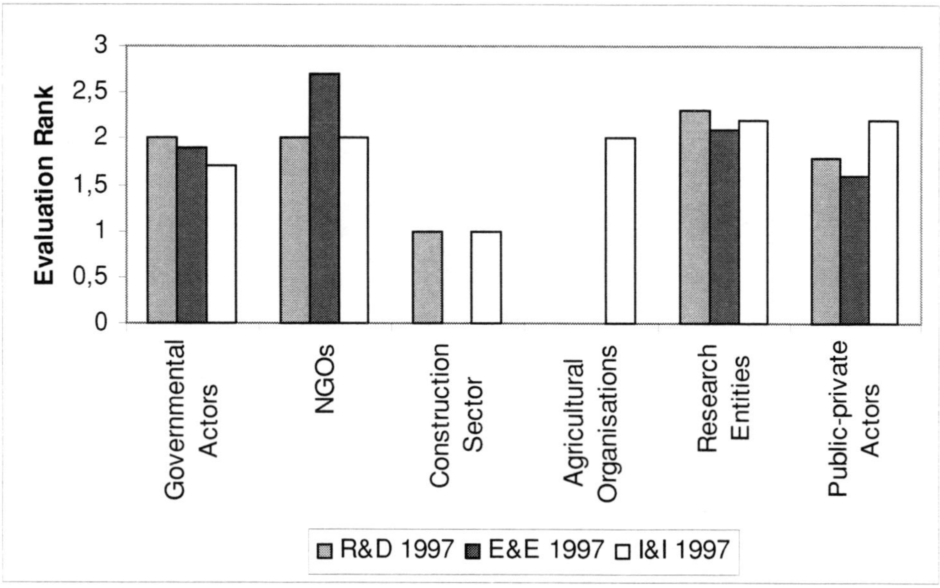

Graphic annex 6.1. Self evaluation of conducted Guadua related activities until 1997 by meso-level actors. Mean evaluation rank per actor group.

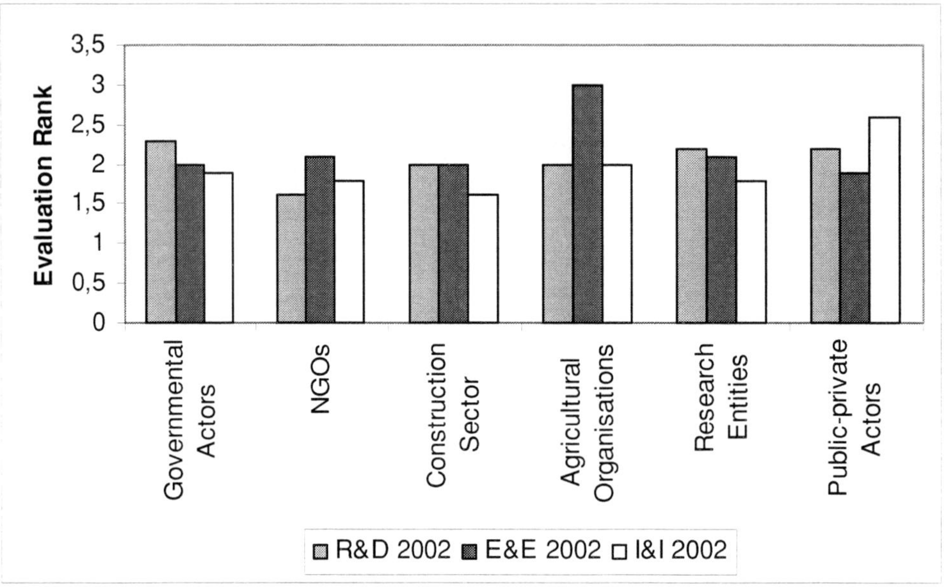

Graphic annex 6.2. Self evaluation of conducted Guadua related activities between 1998 and 2002 by meso-level actors. Mean evaluation rank per actor group.

Annex 7: Correlation Tables for Intensity, Self-evaluation and Co-operations of R&D-, E&E- and I&I-activities

Table annex 7.1. Correlations between intensity, self-evaluation and co-operations in R&D

			Intensity R&D 2002	Evaluation R&D 2002	Average Cooperations R&D
Spearman-Rho	Intensity R&D 2002	Correlation Coefficient	1,000	,618	-,600
		Sig. (2-tailed)	.	,191	,208
		N	6	6	6
	Evaluation R&D 2002	Correlation Coefficient	,618	1,000	,088
		Sig. (2-tailed)	,191	.	,868
		N	6	6	6
	Average Cooperations R&D	Correlation Coefficient	-,600	,088	1,000
		Sig. (2-tailed)	,208	,868	.
		N	6	6	6

Table annex 7.2. Correlations between intensity, self-evaluation and co-operations in E&E

			Intensity E&E 2002	Evaluation E&E 2002	Average Cooperations E&E
Spearman-Rho	Intensity E&E 2002	Correlation Coefficient	1,000	-,736	,486
		Sig. (2-tailed)	.	,096	,329
		N	6	6	6
	Evaluation E&E 2002	Correlation Coefficient	-,736	1,000	-,088
		Sig. (2-tailed)	,096	.	,868
		N	6	6	6
	Average Cooperations E&E	Correlation Coefficient	,486	-,088	1,000
		Sig. (2-tailed)	,329	,868	.
		N	6	6	6

Table annex 7.3. Correlations between intensity, self-evaluation and co-operations in I&I

			Intensity I&I 2002	Evaluation I&I 2002	Average Cooperations I&I
Spearman-Rho	Intensity I&I 2002	Correlation Coefficient	1,000	,667	,145
		Sig. (2-tailed)	.	,148	,784
		N	6	6	6
	Evaluation I&I 2002	Correlation Coefficient	,667	1,000	,265
		Sig. (2-tailed)	,148	.	,612
		N	6	6	6
	Average Cooperations I&I	Correlation Coefficient	,145	,265	1,000
		Sig. (2-tailed)	,784	,612	.
		N	6	6	6

Table annex 7.4. Correlations between average co-operations in R&D, E&E and I&I

			Average Cooperations R&D	Average Cooperations E&E	Average Cooperations I&I
Spearman-Rho	Average Cooperations R&D	Correlation Coefficient	1,000	-,657	,232
		Sig. (2-tailed)	.	,156	,658
		N	6	6	6
	Average Cooperations E&E	Correlation Coefficient	-,657	1,000	-,232
		Sig. (2-tailed)	,156	.	,658
		N	6	6	6
	Average Cooperations I&I	Correlation Coefficient	,232	-,232	1,000
		Sig. (2-tailed)	,658	,658	.
		N	6	6	6

Table annex 7.5. Correlations between intensity in R&D, E&E and I&I

			Intensity R&D 2002	Intensity E&E 2002	Intensity I&I 2002
Spearman-Rho	Intensity R&D 2002	Correlation Coefficient	1,000	,943**	,657
		Sig. (2-tailed)	.	,005	,156
		N	6	6	6
	Intensity E&E 2002	Correlation Coefficient	,943**	1,000	,600
		Sig. (2-tailed)	,005	.	,208
		N	6	6	6
	Intensity I&I 2002	Correlation Coefficient	,657	,600	1,000
		Sig. (2-tailed)	,156	,208	.
		N	6	6	6

**. The correlation is significant at level 0,01 (two-tailed).

Table annex 7.6. Correlations between self-evaluation of R&D, E&E and I&I

			Evaluation R&D 2002	Evaluation E&E 2002	Evaluation I&I 2002
Spearman-Rho	Evaluation R&D 2002	Correlation Coefficient	1,000	-,485	,373
		Sig. (2-tailed)	.	,330	,466
		N	6	6	6
	Evaluation E&E 2002	Correlation Coefficient	-,485	1,000	-,149
		Sig. (2-tailed)	,330	.	,778
		N	6	6	6
	Evaluation I&I 2002	Correlation Coefficient	,373	-,149	1,000
		Sig. (2-tailed)	,466	,778	.
		N	6	6	6

Annex 8: Detailed list of interviewed meso-level actors in 2004

Institutional background	Number	Details
Regional SME promoting organisations	3	• INCUBAR Manizales • INCUBAR Pereira • ANDI Pereira
Actors promoting R&D at national level	2	• COLCIENCIAS • ICONTEC
National Guadua related organisations	2	• MSBC Program • Guadua Productive Chain Board
Banks with SME credit lines	3	• Banco CONAVI • BANCAFE • BANCOLOMBIA
NGOs with SME credit lines	2	• ACTUAR Manizales • ACTUAR Armenia
National organisations of the forestry and wood processing sectors	2	• CONIF • Ministerio del Medio Ambiente: Department for the planning of the national forest plan

Annex 9: Total of COLCIENCIAS funded projects from 1993 to 2003

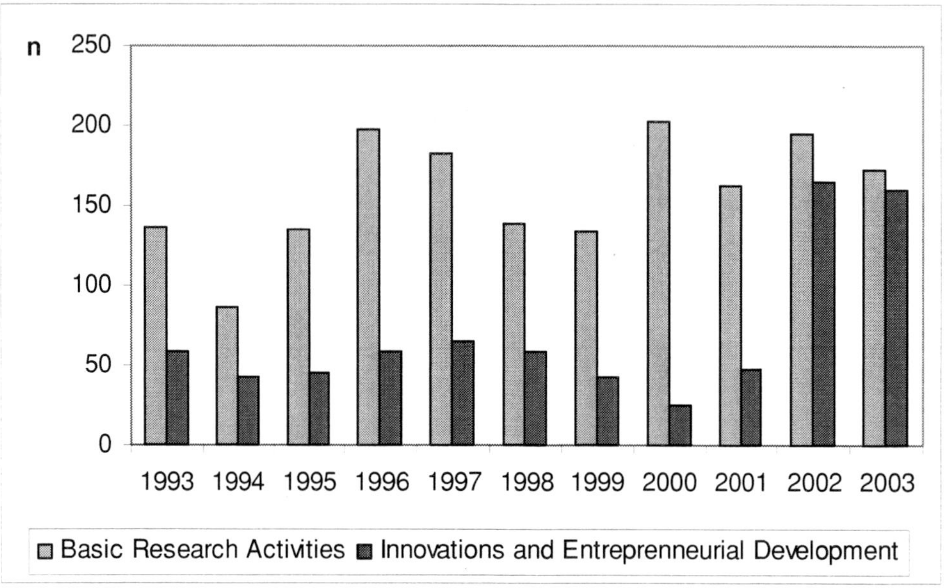

Graphic annex 9.1. Number of COLCIENCIAS funded projects 1993-2003 according to programmatic line

In the *graphic annex 9.1* the increasing importance of innovation and entrepreneur related activities becomes obvious. From the year 2002 on the number of this class of funded projects has almost reached the number of funded basic research projects. This is related to the increase of provided funds for this type of research activities, as it was stated by COLCIENCIAS staff members.

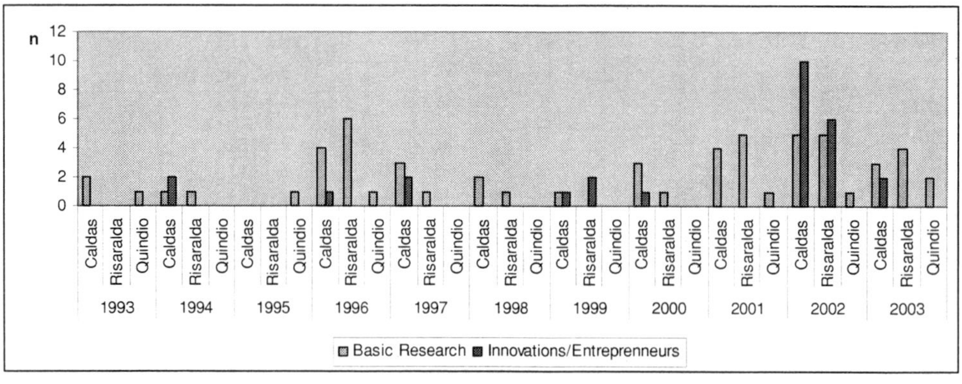

Graphic annex 9.2. COLCIENCIAS funded projects in three departments of the study region between 1993 and 2003